A Mole of Chemistry

An Historical and Conceptual Approach to Fundamental Ideas in Chemistry

Caroline Desgranges and Jerome Delhommelle

CRC Press
Taylor & Francis Group
Boca Raton London New York

CRC Press is an imprint of the
Taylor & Francis Group, an **informa** business

First edition published 2020

by CRC Press
6000 Broken Sound Parkway NW, Suite 300,
Boca Raton, FL 33487-2742

and by CRC Press
2 Park Square, Milton Park, Abingdon, Oxon, OX14 4RN

© 2020 Taylor & Francis Group, LLC
CRC Press is an imprint of Taylor & Francis Group, LLC

International Standard Book Number-13: 978-0-367-20828-8 (Hardback)
International Standard Book Number-13: 978-0-367-20824-0 (Paperback)
International Standard Book Number-13: 978-0-429-26368-2 (eBook)

Typeset in Times
by Deanta Global Publishing, Services, Chennai, India.

Cover Image: Arthur Schuster and Ernest Rutherford with students: Physics Department, Manchester
University. Photograph. Credit: Wellcome Collection. CC BY

Contents

Authors

Caroline Desgranges, PhD, earned a DEA in physics in 2005 at the University Paul Sabatier – Toulouse III (France) and a PhD in chemical engineering at the University of South Carolina in 2008.

Jerome Delhommelle, PhD, earned a PhD in chemistry at the University of Paris XI-Orsay (France) in 2000. He is a former student of the Ecole Normale Superieure at Cachan (France) and was awarded the Agregation in Physical Sciences (Chemistry track) in 1997. He is currently an Associate Professor in chemistry at the University of North Dakota.

Introduction to Modern Chemistry
The Beginning of Scientific Reasoning

"Gedanken ohne Inhalt sind leer, Anschauungen ohne Begriffe sind blind".

Kant, *"Kritik der reinen Vernunft"*, 1781

"Thoughts without content are empty; intuitions without concepts are blind".

Kant, "Critique of Pure Reason", 1781

THE CENTURY OF PHILOSOPHY OR THE AGE OF REASON

It takes two centuries and several revolutions to see a part of natural philosophy (the study of Nature) transformed into what we now call modern chemistry. The movement begins with the Copernican revolution (16th century), where science is now defined around core values, such as reflection, deduction and reasoning supported by experimental data. The discoveries by Galileo further establish heliocentrism, overthrowing the established order and, in particular, the vision of the world, the concepts of matter and life, previously conceived by religions. The Age of Enlightenment is a turning point, as knowledge is not based anymore simply on dogma, but necessarily relies on critical thinking. Descartes, in his "Discourse on Method" (1637), gives the foundations of scientific reasoning by basing its principle on doubt and on the *cogito* ("I think therefore I am"). Many philosophers work and propose different concepts to better understand, but also analyze our thinking: Spinoza ("On the Improvement of the Understanding", 1677) and Locke ("An Essay Concerning Human Understanding", 1689) are just some examples of an impressive movement. They define and explore how we build our ideas and thoughts with respect to our background (religion, origin, social class, etc.) and how this influences our representation of what surrounds us. More specifically, they study how rationalism (reflection based on mathematical reasoning) and empiricism (reflection based on experimental facts) can advance human knowledge. It is an exciting time during which Newton and Leibniz invent differential calculus (giving rise to calculations of velocity and acceleration) and integral calculus (allowing the understanding of the world in 2D and 3D with calculations of area and volume). In 1687, Newton publishes his *"Philosophiae Naturalis Principia Mathematica"* that will remain a reference for all scientists until the 20th century. Let us also note the significance of Kant's work (18th century) focusing on the limitations of our own reason to critique itself ("Critique of Pure Reason", "Critique of Practical Reason", "The Metaphysics of Morals", "Critique of Judgment"). It is during the Age of Enlightenment that numerous philosophers try to determine either the subjective or the objective truth behind the phenomena surrounding them, and, in turn, to understand our interactions with Nature.

THE RISE OF EXPERIMENTAL CHEMISTRY

Starting from the 18th century, experimental chemistry rapidly expands. Important scientific discoveries allow for the development of new theories which, for the most part, remain valid to this day. Among these, let us note the pioneering work of Priestley on gases, which later results in a new kind of chemistry known as pneumatic chemistry (chemistry of gases). He is also one of the first scientists who chooses to work from facts that are derived from experiments, rather than to state new theories without having any experimental evidence to start from. At this time, only the air, fixed air (carbon dioxide – Black, 1754) and inflammable air (hydrogen – Cavendish, 1766) are known. Between 1774 and 1786, Priestley publishes "Experiments and Observations on Different Kinds of Air" and sheds light on numerous "new" gases, through his experiments on combustion using what is called a "burning lens". This experimental device is composed of a convex lens that allows the focusing of sunlight. Such an idea dates back to Archimedes and his so-called "heat ray" which would have, according to the legend, destroyed the Roman fleet during the siege of Syracuse in 213 BC. During his work, Priestley discovers nitric oxide, nitrogen dioxide, nitrous oxide, ammonia, sulfur dioxide, hydrogen chloride, nitrogen and carbon monoxide as well as oxygen, also isolated by Scheele around that time. Most notably, he proves the existence of oxygen by carrying out a series of simple experiments. He lights a candle and places it under a glass jar. He discovers that the candle goes out after a while, deducing that there was a different kind of air present under the jar when the candle was burning. This brings him to identify the "dephlogisticated air" now known as (oxygen) that allows the candle to remain lit. Indeed, at that time, the combustion process was understood through the phlogiston theory, introduced by Becher and Stahl at the end of the 17th century. According to this theory, when charcoal burns, it releases an amount of phlogiston (Greek: fire of the Earth) into the surrounding air, yielding phlogisticated air. Priestley is the first to write the chemical equation related to this phenomenon as follows: charcoal (containing phlogiston: $+\Pi$) + air = fixed air (CO_2) + phlogisticated air. Similarly, he is able to study the reaction of mercury with oxygen through the following equation: mercury ($+\Pi$) + dephlogisticated air (air without phlogiston) = calx of mercury. We will see later that this theory will be replaced by the caloric theory, opening the door to a new field that will become thermodynamics (the study of heat). Later, Priestley repeats the experiment with the candle, but places mint next to the candle inside the jar. He discovers that the candle still goes out after a while. However, to his great surprise, after a few days, he finds that the candle can burn again, demonstrating that the presence of mint has produced dephlogisticated air! This is an amazing fact that will allow for the discovery of photosynthesis in the 19th century. He also studies the breathing process by placing mint and a mouse next to each other under the bell.

THE CHEMICAL REVOLUTION

One of the biggest discoveries of the 18th century is Lavoisier's law of conservation of mass. It is based on a series of experiments begun in 1772, for which Lavoisier uses a highly accurate balance. He finds that when inflammable air reacts with dephlogisticated air to give water, there is no mass present except that of the reactants and products of the reaction, meaning there is no sign of phlogiston. He follows the same process for different chemical reactions and finds that, in every experiment he carries out, there is no mass to account for phlogiston! He deduces that *"Rien ne se perd, rien ne se crée, tout se transforme"* (Nothing is lost, nothing is created, everything is transformed). From this series of experiments, he also shows that the air is composed of two gases. The first one, oxygen (Greek: acid former), is responsible for combustion and acidity. The other one, named azote (Greek: without life), is now known as nitrogen. Other famous experiments deal with the combustion of phosphorus and sulfur. In 1789, he publishes *"Traité élémentaire de chimie"*, now regarded as the first modern chemistry textbook. He formalizes his discoveries by introducing the concepts of elements (the simplest substance that can be extracted from other substances) and of chemical equations (mathematical formulae relating the various elements present in the experiment). All this is governed by the law of conservation of mass, which underlies chemical reactions. Lavoisier's law is one of the most important principles in chemistry. It forever removes the magic character of alchemy from chemistry. It also finally provides a rational framework and basis for modern chemistry. Thanks to his experiments on combustion, Lavoisier is able, at the end of the 18th century, in *"Méthode de nomenclature chimique"* to give the first classification of elements in which oxygen, azote, hydrogen, sulfur, phosphorus, charcoal, gold, platinum, mercury, iron, tin, zinc, antimony, silver, arsenic, bismuth, cobalt, copper, manganese, molybdenum, nickel, lead and tungsten all already appear together with light and caloric!

FIRST ATOMIC THEORIES

In 1808, Dalton publishes "A New System of Chemical Philosophy", in which he proposes the first atomic theory. It is mostly based on his pneumatic experiments and, particularly, on his law on partial pressures. Indeed, he demonstrates that the total pressure of a gas in a given volume can also be defined as the sum of the partial pressures of each component of this gas, taken separately in the same volume (Dalton's law, 1801). It brings to light the fact that matter is composed of subsets, and more specifically, that each element is composed of atoms. The atomic theory retains the basic principle stated by Ancient Greeks that every element is composed of atoms, which are small indivisible particles. He also defines that all atoms of the same element are identical, and that atoms differ from one element to the other. Finally, he states that it is the rearrangement, the separation and the combination of atoms which are at the center of the mechanisms taking place during a chemical reaction. In short, according to Dalton, matter consists of atoms of different masses that are combined in simple proportions. This theory echoes the ideas of Ancient Greeks that atoms love or hate each other (concepts of attraction and repulsion). Similar ideas yield significant developments in the field of electricity at the beginning of the 19th century. Experimentalists use electric decomposition, or electrolysis, to show that two identical atoms, both carrying a positive charge, repel each other, whereas atoms with opposite charges attract one another, allowing for bond formation. One of the consequences of this atomic theory is that two hydrogen atoms cannot be put together to create a molecule. A water molecule is therefore written as OH and ammonia as NH. Dalton also uses the concept of equivalent weights, discovered by Wenzel and Richter at the end of the 18th century, to finalize his atomic theory, and begins the first classification of elements. One of his hypotheses is to take hydrogen as the reference for the unit mass. For example, he determines the equivalent mass of oxygen through the following experiment. In the case of steam (gaseous water) formation, it was known at that time that 1 g of hydrogen reacting with 8 g of oxygen gives 9 g of steam. From this observation, and according to the atomic theory that states H + O → HO, he deducts that the equivalent weight of oxygen is 8 g. Nowadays, the Dalton is the standard unit for the measure of the atomic mass, but rather than being defined relative to hydrogen, it is defined as 1/12 of the mass of a carbon atom for the ^{12}C, and we shall see later why.

STOICHIOMETRY

Berzelius is one of the founders of modern chemistry, working at the beginning of the 19th century. He demonstrates that chemical proportions, or stoichiometry (the term introduced by Richter in 1792), allow the analysis of the quantities of reactants and products during a chemical reaction. His idea is to use this concept to define the proportion of an element in a chemical formula. For this purpose, he determines the number of simple atoms that compose a more complex substance, as well as their relative weights. In 1814, he publishes his first table of atomic weights. This table differs from the one proposed by Dalton, as it takes oxygen as the reference. Berzelius also introduces the chemical notation in which the composition of molecules is indicated, as in mathematics, with an exponent. A water molecule would have been written H^2O at this time. He establishes the use of Latin to name chemical elements: Pb for plumbum (English: lead). He also discovers new elements such as cerium (1803), selenium (1818) and thorium (1823).

ATOMS AND MOLECULES

Another very significant advance is the distinction between atoms and molecules. The beginning of an answer is proposed by Avogadro, following the work of Gay-Lussac on the proportions of volumes. In his experiments, Gay-Lussac demonstrates that a substance like steam is actually composed of two volumes of hydrogen gas and one volume of oxygen gas. This allows him to write the chemical formula for water, now known as H_2O. However, recall that at the time, this discovery contradicts Dalton's work and his atomic theory that defines a water molecule as HO. Avogadro suggests a new hypothesis, based on Gay-Lussac's work, that equal volumes of gas, at the same temperature and same pressure, always contain the same number of molecules. This is fundamental, since this is the first time that molecules and atoms are defined.

It means that, in the case of water, two volumes of hydrogen gas and one volume of oxygen gas are needed to obtain two volumes of steam. It also means that, in each of these volumes, there is the same number of molecules of gas, regardless of the nature of the gas considered. Building on the law of conservation of mass (Lavoisier), we can thus determine the number of molecules, as well as the number of atoms, in any volume of gas. To illustrate this point, we start from Dalton's formula for water: $H + O \rightarrow HO$. Adding Gay-Lussac's observations, we end up with $2H + O \rightarrow 2HO$. This is incorrect, since it doesn't obey the law of conservation of mass (1O on the reactant side compared with 2O on the product side)! We need a new formula for water! If we start over from Avogadro's hypothesis combined with Gay-Lussac's finding, we obtain the following: two volumes of hydrogen (containing one hydrogen molecule in each volume) + one volume of oxygen (containing one oxygen molecule) $\rightarrow$ two volumes of steam (containing one water molecule in each volume). Following Lavoisier's law, the reaction should obey the law of conservation of mass. If one hydrogen molecule is composed of one atom and one oxygen molecule is composed of 1 atom, the total number of atoms on the reactant side is three atoms ($2 \times 1H + 1 \times 1O$). This tells us that, on the product side, we should have a total of three atoms for two molecules of water. This is not possible since 3/2 is not a whole number. This would mean that there are 1.5 atoms per water molecule, which is impossible since atoms are indivisible. On the other hand, if we now make the assumption that one hydrogen molecule is composed of two atoms and assume the same for the oxygen molecule, we have six atoms on the reactant side. This means that each of the two water molecules contains three atoms (6/2). We end up with a new formula for water: H_2O! Thus, we have $2H_2 + O_2 \rightarrow 2H_2O$. Oxygen gas is therefore composed of two oxygen atoms, which is why we name it dioxygen (O_2). This is also the case for dihydrogen (H_2). We add that most gases are diatomic.

AVOGADRO'S NUMBER

Facing the recurrent issue of the distinction between molecules and atoms (as we saw, most gases are diatomic molecules), renowned scientists, including Kekulé and Wurtz, gathered the foremost chemists to solve this problem. This is how the first world conference of chemistry is organized in 1860. It is also known under the name of Karlsruhe congress, from the name of the city in Germany where it is held. It is on this occasion that Cannizzaro, Avogadro's former student, discusses new rules allowing for the calculation of molecular weights from atomic weights. He uses the density in g/cm^3 or kg/L (fraction of the mass of a given substance over the volume V occupied) if the substance in its gaseous form at 0°C and at 1 atm, as well as the mass percentage of each element composing the substance studied. For example, ammonia (NH_3) is composed of nitrogen and hydrogen and has a density of 0.76 g/L. Its molar mass can be calculated from Avogadro's law using the following basis: 1 mole of ideal gas occupies a volume of 22.4 L. This gives a molar mass for ammonia of 22.4×0.76 = 17 g/mol (1 mole of ammonia weighs 17 g). Moreover, ammonia contains a mass percentage of 17.6% of hydrogen which gives $17 \times (17.6/100) = 3$ g/mol of hydrogen. In other words, there are 3 g of hydrogen in 1 mole of ammonia. Finally, we conclude that, since a mole of ammonia weighs 17 g and there are 3 g of hydrogen in 1 mole of ammonia, there is also 14 g of nitrogen per mole of ammonia ($17 - 3 = 14$ g). This means that the molar mass for nitrogen is 14 g/mol. Unfortunately, many of the elements that we know are not in a gaseous form at ambient temperature and pressure, and thus, these rules do not apply to all elements. It is the use of calorific capacities (law of Dulong and Petit) that will allow for the calculation of the molar mass for the other elements. From there, chemistry will make spectacular progress in the following years, with, for example, the publication by Mendeleev of the famous periodic table. We add that the well-known Avogadro's number (N_a) was named in recognition of Avogadro's work. Based on the kinetic theory of gases, Loschmidt will find, in 1865, that the order of magnitude of this number is 10^{25}, which, in turn, means that a molecule has a size of about 1 nm (10^{-7} cm). It is only in 1961 that the International Union of Pure and Applied Chemistry (IUPAC) fixes the definition that we now use for atomic mass as 1/12 of one atom of ^{12}C, that is, 12 grams of ^{12}C contains N_a atoms. Today, thanks to the amazing accuracy of experimental devices, we can calculate that $N_a = 6.022 \times 10^{23}$ mol^{-1}.

1 Introduction to the Kinetic Theory of Gases
The Ideal Gas Law

"Sed quid igitur sum? Res cogitans. Quid est hoc? Nempe dubitans, intelligens, affirmans, negans, volens, nolens, imaginans quoque, & sentiens".

Descartes, "*Meditationes de Prima Philosophia*", 1641

"But what then am I? A thing that thinks. What is that? A thing that doubts, understands, affirms, denies, wills, refuses and that also imagines and senses".

Descartes, "Meditations on First Philosophy", 1641

A WILD SPIRIT NAMED GAS

One of the most interesting concepts in science is certainly the notion of gas. Even if, etymologically, the term does not appear until the 17th century, it is already present in another form during antiquity. Indeed, "Χάος" (Greek: Chaos), which translates as "wide open chasm" in ancient Greek, is found at the origin of the world and the gods in the Theogony (Greek: birth of the gods) of Hesiod (8th century BC). Remember that during the Greek antiquity, epics and poems are sung by aoidos (Hesiod being one of the most famous), who are widely acclaimed in the cities they tour. The format of this poetry is highly codified. It is based on dactylic hexameters, the most common form in Greek epics, like the "Iliad" and the "Odyssey" by Homer (another famous aoidos whose existence is still a mystery). According to Greek mythology (see Figure 1.1), Chaos gives birth to the Earth (Gaia), which in turn engenders the Sky (Uranus), the Mountains (Ourea) and the Sea (Pontus). From these follow the four elements, pillars of Aristotle's thought (384–322 BC). Gods are represented by men or women, endowed with great powers that make them immortal gods, and who, by their interactions, control the Universe. An interesting part of Theogony is that Chaos' descendants also include darkness (Erebus), night (Nyx), day (Hemera) and upper sky (Aether). Note that the latter is the air that gods breathe, unlike Aer, the air from the lower parts of the sky, that mortals breathe. Still in Theogony, Erebus, who helps the Titans (primordial giant divinities preceding the gods of Olympus) during their war with Zeus, is sent deep beneath the surface of the Earth. Erebus then becomes the name of a region of the Underworld, where the souls of the dead pass on their way to Hades (god of the dead). We can already see two definitions emerge: Chaos as the origin of a specific air, and Chaos as related to the soul of humans. Rather an interesting concept since it starts from nothing.

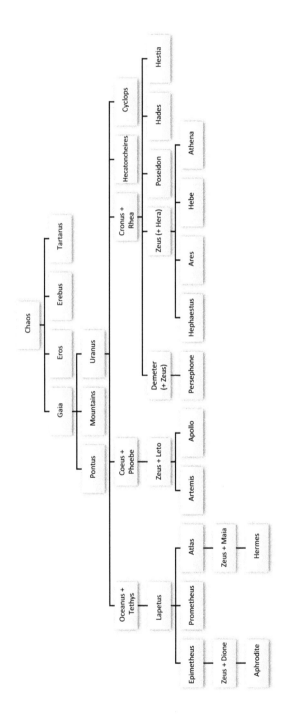

FIGURE 1.1 Theogony – Greek mythology family tree.

Later, Ovid (43 BC–17 AD), one of the most famous Latin poets, gives the following description of Chaos in the "Metamorphoses":

Ante mare et terras et quod tegit omnia caelum
Vnus erat toto naturae vultus in orbe,
Quem dixere chaos: rudis indigestaque moles
Nec quicquam nisi pondus iners congestaque eodem
Non bene iunctarum Discordia semina rerum.

Before the ocean and the Earth appeared – before the skies had
overspread them all – the face of Nature in a vast expanse was naught
but chaos, uniformly waste. It was a rude and undeveloped mass, that
nothing made except a ponderous weight; and all discordant elements
confused, were there congested in a shapeless heap.

Let's add that this concept of chaos remains unchanged during the Imperium
Romanum (Latin: Roman Empire) which extends from Western Europe to Asia
Minor, accounting for a fifth of the entire world population at the time! It is only in
476 AD (fall of Rome) for the western part and 1453 (fall of Constantinople) for the
eastern part that the empire collapses.

A new paradigm appears with the advent of alchemy, it seems during the 5th–
6th centuries AD according to Greek literature. For some, it comes from the word
χημεία (khemeia) translated into Arabic in الكيمياء (al-kimiya) and translated
into Latin during the Middle Ages in alchemia. For others, it would come from
⌂𓈖𓏤 (km•t: Egypt, black land), translated into Coptic and then into Greek (see
Rosetta Stone). The main idea of the alchemists is to discover the laws of Nature
that surrounds us and eventually control it. This includes the transmutation of metals
thanks to the philosopher's stone, the search for the panacea and the extension of life
via an elixir of long life. The idea is not to become rich or immortal, but to demonstrate, with a few practical examples such as transmutation, that alchemists have
reached a total understanding of Nature. To achieve this, they need a theory that will
take the shape of a natural philosophy based on experiments. Indeed, the word laboratory is invented by alchemists during the 17th century! Their work, like a building
game, aims at understanding Nature from the materials it is made of (see Figure 1.2).
The first alchemy books appear during the 11th century, in the form of practical
treatises dealing with dyeing fabrics and metalwork. According to alchemists, each
material is made of three things, called "fundamental principles": mercury, sulfur
and salt, the famous alchemical trinity! For example, when a piece of wood burns,
it is the sulfur it contains that burns, the mercury it contains that is carried away by
the smoke and the salt it contains that remains in the ashes. In other words, salt represents the principle of fixity (the part of wood that remains after combustion) which
serves as a basic constituent for each thing. Mercury, the principle of life (ability
for the wood to change shape during combustion) which allows matter, things, to
melt and flow, processes transcending the solid and liquid states of matter. It is also
known as the "mind" of things. Finally, sulfur, the principle of change, is the driving
force or impetus for things to change, giving, for instance, wood its inflammability
and volatility. For alchemists, it represents the "soul" of things. The idea is that any

substance can be decomposed and recombined to form another substance! Applied to metals, it simply means that a common metal could be transformed into gold (the famous transmutation process)! Let us add that the concept of chaos can be related to the notion of "*Prima Materia*", as described by Ruland in "*Lexicon Alchemiae*" (1612), "A crude mixture of matter or another name for Materia Prima is Chaos, as it is in the Beginning". At the other end of the alchemical spectrum, we find Paracelsus (1493–1541) who practices an alchemical medicine. Indeed, he invents "spagyric", the art of separating and assembling the main constituents of bodies for medical, as well as philosophical, purposes. Indeed, he is the first to come up with the idea that a human being is a complex combination of a visible body and of an invisible soul that can both have a profound impact on their health. Let us note the publication, during the Renaissance, of numerous treatises discussing the spirituality of Nature, often through the lenses of mystical constructions. Just remember that the most popular experimental apparatus of that time is the alembic, which can turn a liquid into a "spirit"! It is finally van Helmont (1579–1644), an alchemist from the 17th century who, during one of his combustion experiments, introduces the word gas in "*Ortus Medicinae*": "Suppose thou, that 62 pounds of Oaken coal, one pound of ashes is composed. Therefore the 61 remaining pounds are the 'wild spirit' which, also being

FIGURE 1.2   Names and symbols used in alchemy in "*Traite de la Chymie*" by Le Fevre (1669).

fired, cannot depart, the vessel being shut. I call this spirit, unknown hitherto, the new name of 'gas', which can neither be retained in vessels nor reduced to a visible form, unless the seed is first extinguished." We add that van Helmont is also the first to identify the "gas sylvester" (Latin: *silvestris* meaning wild or wooded – nowadays known as carbon dioxide). Finally, remember that one of the most famous alchemists is Isaac Newton. His laboratory notebooks have now been digitized and made available to the public!

This scientific spirituality brings us back to the modern definition of spirituality for humankind as awakened consciousness. In trying to better understand Nature around us, we are also reaching an inherent duality. As we have seen, mythology provides for a personification of "nothingness" that cannot exist as Aristotle claims (Nature abhors a vacuum). For instance, if high exists, then low must also exist, so that balance is achieved in the world. Later, this line of questioning will give rise to the advent of metaphysics. According to Descartes (1596–1650) ("Discourse on the Method"), metaphysics deals with God, our minds and "all the simple and clear notions which are in us", i.e. our ideas. Unlike physics, which focuses on material objects, metaphysics is therefore characterized as the knowledge of the immaterial. Therefore, the famous phrase: "I think therefore I am". Thinking is immaterial, which doesn't prevent us from existing! For Descartes, philosophy can be represented by a tree whose roots are metaphysics, the trunk is physics and the branches are mechanics, morals and medicine. Thus, no one can bring any meaningful answers in the last three sciences (or pick the fruits of the tree) unless one has first answered the fundamental questions of metaphysics (or build knowledge on secure foundations). For Spinoza (1632–1677), metaphysics represents a conception of God and the world. In his major book "Ethics", he identifies that human beings always want to know the "why" and constantly strive to give meaning to the world, to the natural phenomena and to their existence. According to him, "men act always on account of an end, viz. on account of their advantage, which they want. Hence they seek to know only the final causes of what has been done, and when they have heard them, they are satisfied, because they have no reason to doubt further."

From a scientific standpoint, it is Lavoisier who finally introduces the word "gas" as a state of matter in his experiments on combustion: "there are three states of matter: the state of solidity, the state of liquidity and the aeriform state [...] I will henceforth designate these aeriform fluids under the generic name of gas" ("Elements of Chemistry", 1789).

FOUNTAIN STORIES IN FLORENCE, ITALY

It was during the Renaissance that the question of the existence (or not) of vacuum resurfaced. Indeed, it is in the context of the Italian Renaissance in Florence that everything unfolds. At that time, Florence is one of the most important cities in Europe. Florence is regarded as the cradle of the Renaissance, as a result of its tremendous influence in terms of culture, arts and sciences. The Medici, rich merchants and then bankers, participate from generation to generation in its growth. Indeed, through patronage, they become more and more involved in the life of the city to finally emerge as its leaders for more than four centuries (from the 14th to the 18th century). The 15th century is the golden age of Florence, during the reign of Lorenzo de Medici (1449–1492), known as Lorenzo "the Magnificent". Inspired by Machiavelli (1469–1527) (who some consider as the father of political science) and his work "*Il Principe*" ("The Prince"), Lorenzo is often regarded as one of the greatest political and cultural minds in Italy at the end of the 15th century. To strengthen his power, he spends vast amounts of money to commission the best artists to create the most beautiful monuments, including Michelangelo, Leonardo da Vinci and Raffaello!

After Lorenzo's death, the Medici dynasty falters. It is only in 1569 that a younger branch of the Medici family regains power in Florence. Indeed, Cosimo I (1519–1574) becomes the first Grand Duke of Tuscany! Always seeking to accrue power through the arts, the Medici family buys Palazzo Pitti and makes it the official residence of the ruling family of the Grand Duchy of Tuscany. The Boboli Gardens are a perfect example of this demonstration of power over Nature through art and science (see Figure 1.3). These gardens of the Italian Renaissance are indeed the model that French-style gardens, such as those of the Chateau de Versailles, will later follow. Their design is based on geometry, forming symmetrical patterns with perspective. Fountains and waterfalls have the aim of enlivening the gardens while evoking the prosperous period of Ancient Rome. Overall, the goal is to create harmony and order, key concepts from the Renaissance. However, the Boboli Gardens lack natural water springs. Following Cosimo's ideas, fountains and aqueducts are built, providing a source of drinking water to the Florence population, but also serving to reaffirm his power. For instance, he asks Ammannati (1511–1592) to build the famous fountain of Neptune which is the first fountain to operate continuously! Let us add that Ammannati portrays Neptune as Cosimo to celebrate Cosimo's power. The fountains thus become central elements of the town squares, but also of the new Italian Renaissance gardens. So, it is not a surprise when Ferdinando II de Medici

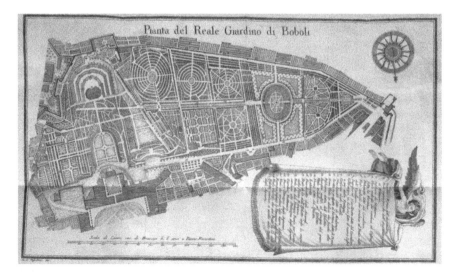

FIGURE 1.3   18th century map of the "*Giardino di Boboli*" (Boboli gardens).

decides to enlarge the Boboli Gardens during the 17th century to reach the area of 11 acres (45,000 m²) that we know today.

At that time, the fountain engineers of Florence start building gigantic hydraulic installations in the gardens of the palace. They install suction pumps but discover, to their great amazement, that they are unable to raise water beyond 33.9 feet (10.33 m), as if there were a limit. Facing this challenge, they call upon the best scientist in the city: Galileo (1564–1642). Let us recall that, after losing his trial against the Inquisition in his bid to defend the model of Copernicus on heliocentrism (from which follows the famous phrase "And yet it moves"), Galileo returns to Florence, old and blind. Unfortunately, Galileo dies in 1642 without officially giving a solution to the famous fountain enigma. Nevertheless, we will find later in his notes that he had thought that the air should weigh something, notes that, perhaps, Torricelli (1608–1647), his successor at the Florentine Academy, has read. Torricelli is the one who finds the answer to this fascinating puzzle! The first problem he encounters is the actual size of the experimental apparatus. Indeed, studying something about 10 meters high is not easy. His genius idea is to replace water with mercury (also known as quicksilver). Indeed, we know that mercury is more dense (or in other words, heavier) than water. To verify this, if one weighs a liter of water, one finds it to be 1 kg (thus the density of water is 1 kg/L or, in other units, 1 g/cm³). On the other hand, if one weighs 1 liter of mercury, one finds it to be 13.56 kg (hence a density of 13.56 kg/L or 13.56 g/cm³ for mercury).

To get back to our fountains, the basic idea is that they operate using a suction pump, which raises a column of water from the Arno river to the garden (but only up to a certain limit, the famous 10.33 m). According to Aristotle, since "Nature abhors a vacuum", a space that is initially empty always fills up. In our case, it means that water sticks to the air sucked out by the pump. Therefore, the famous limit was explained by the possible break of the water column. However, Torricelli does not think there is such a force inside the tube of the pump (called a siphon), but that it is the weight of the air, outside of the siphon, that raises the water column! His idea is that the water column of 10.33 m has the same weight as a column of air extending to the top of the Earth's atmosphere. So, if the weight of the air column is greater than the weight of water inside the siphon, the air pushes water beyond the top of the tube and water flows. Otherwise, if the weight of the air column is less than the weight of water, the air cannot push the water high enough (up to the top of the tube) and water cannot flow. To validate his hypothesis, Torricelli takes a tube filled with mercury, closes it with his finger and turns it upside down into a container full of mercury. Since his hypothesis is that air has a weight, it will thus push mercury and create a vacuum. According to his calculations, he finds that if density is the most important parameter, mercury should rise to a height 13.56 times lower than water, i.e. 33.9 feet/13.56, amounting to 2.5 feet or about 760 mm. When Torricelli carries out the experiment, he notices that mercury does not stick to the top of the tube, but stabilizes at a height of 2.5 feet, as expected! Moreover, he observes that there is an empty space at the top of the tube (see Figure 1.4). He also realizes that the height of mercury varies from one day to the next, and he attributes this to changes in the weather … he has just invented the barometer! Nevertheless, Torricelli does not publish his results, fearing for his life. Just remember that Torricelli was one of

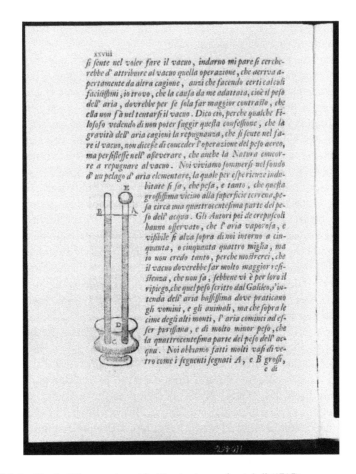

FIGURE 1.4   Torricelli's experiment in *"Lezioni accademiche"* (1715).

Galileo's disciples and that the fear of the Inquisition was deeply entrenched in him. Finally, let us add that Torricelli will publish, three years before his death, *"Opera Geometrica"*, in which he discusses the motion of projectiles and fluids, but nothing on vacuum.

EXISTENCE OF VACUUM AND ATMOSPHERIC PRESSURE

Despite this, the news that there is an experiment suggesting the existence of vacuums starts to spread. Indeed, even if there are no scientific journals as we now have, this does not prevent scientists from exchanging ideas, for instance through letters they send to each other. One of the most important personages of the time is certainly Mersenne (1588–1648), a French theologian. From his austere monastic cell, he maintains a great epistolary correspondence with a multitude of intellectuals all over Europe, including Descartes, Fermat, van Helmont, Torricelli, Huygens, Pascal and Galileo to name a few. He has a keen scientific mind, with broad interests ranging from mathematics to acoustics. Indeed, he writes about the vibrations of strings in

"*Harmonicorum Libri*" (1636) following his discussions with scientists and musicians of the time. It is a wonderful book that, nowadays, makes Mersenne one of the fathers of acoustics (the science behind sounds). He is also a brilliant mathematician who studies prime numbers (for instance 2, 3, 5, 7, 11, 13, 17, 19, 23, 29, 31…). He observes a strange pattern. Consider the prime number 3; it can be written as 4 − 1 or also as $2^2 − 1$. Now consider 7; it is equal to 8 − 1 or $2^3 − 1$. The same is true with 31 which is also 32 − 1 or $2^5 − 1$. As you can see a formula emerges: The form $2^n − 1$ gives a prime number if n is prime itself! Nowadays, we call them the Mersenne numbers M_n (n being a prime number). For example, the first Mersenne number is related to the first prime (2) and we have then $M_2 = 2^2 − 1 = 3$. The same is true with the second Mersenne number (using the second prime: 3) $M_3 = 2^3 − 1 = 7$, ditto with $M_5 = 2^5 − 1 = 31$, $M_7 = 2^7 − 1 = 127$. Note that, as of December 2018, 51 Mersenne numbers are known, the biggest being $M_{82,589,933}$ which is also the largest known prime number! Like many of the other Mersenne numbers, it was discovered by a calculus distributed under the aegis of the Great Internet Mersenne Prime Search (GIMPS) project.

During a trip to Italy, Mersenne witnesses Torricelli's experiment. He is so impressed that he shares it with his correspondents starting in 1645, and, more specifically, with another important scientist: Pascal (1623–1662). Despite his young age, Pascal has already invented the famous Pascaline (one of the first calculators!). He redoes Torricelli's experiment but, this time, he uses water and wine. Indeed, one of the questions that arises is what really happens at the top of the tube. For some, there would not be a vacuum, but a vapor. In order to clarify this, he performs the so-called "liquor experiment". For this, he recreates Torricelli's experiment using water as a reference for his findings. Indeed, he observes that the height of the water column reaches 33.9 feet, leaving an empty space between the surface of water and the top of the tube. Using wine instead of water, he finds that the wine column rises higher than water, leaving a smaller empty space at the top! Let us not forget that this result goes against the physical intuition of the time. Since wine has more "spirits" than water, it should be more volatile and thus the space at the top should be larger than for water … but this is not the case! And Pascal understands why … it is again a density matter. As wine is less dense than water, the wine column will rise higher than the water column – as with mercury and water in Torricelli's experiment. It seems that Pascal's birth in a hilly part of France then leads to his greatest intuition. Indeed, to demonstrate that the height of mercury depends on the weight of air above the apparatus, he plans a series of measurements at different altitudes. This is his famous experiment at Puy de Dome in 1648. To do this, he asks his brother-in-law (Périer) to take with him a barometer and note the height of mercury at different stages of his ascent of the Puy de Dome mountain. At the base of the mountain, Périer and his friends measure 28 inches. Once at the top, they find 25 inches. The experiment shows that Pascal is right! The higher they climb the lower mercury rises into the tube! This means that the weight of air decreases as they go from the base to the top of the mountain! Let us add that, excited by the result, Pascal chooses to redo the experiment, but this time in Paris at la Tour Saint-Jacques (Saint-Jacques Tower). He finds again that the height of mercury decreases when he climbs up the tower and that there is a vacuum at the top of the tube. He will devote an entire book on the topic of vacuum called "*Traité du vide*" in 1651.

Thanks to his experiment at Puy de Dome, Pascal proves the existence of a pressure, from the Latin *pressio* (action of weighting), exerted by the air on mercury. Nowadays, the notion of pressure is often used to measure height (or in other words, as a vertical coordinate in mountaineering or aeronautics). As a tribute to the work of Pascal and Torricelli, two units for pressure were created: Pa (for Pascal) and Torr (for Torricelli). One Torr corresponds to 1 mmHg (1 millimeter of mercury). Using Torricelli's experiment, under ambient conditions, the height of mercury reaches 30 inches, or 760 mmHg, which is equivalent to a pressure of 1 atm (i.e. the atmospheric pressure). Let us add that another unit for pressure is also commonly used, the bar, with 1 bar = 100,000 Pa or 10^5 Pa.

One of the most impressive events in the history of pneumatic chemistry is certainly the well-known "Magdeburg hemispheres" experiment directed by Otto von Guericke (1602–1686) (see Figure 1.5). In 1654, in front of Emperor Ferdinand III and the princes, two copper hemispheres of 14 inches (35.5 cm) in diameter are brought together at Regensburg (now in Germany). The system is maintained with a mixture of grease, wax and turpentine. An important detail for the future: one of the spheres has a tube closed by a valve. The show begins and von Guericke asks that two teams of eight horses separate the hemispheres. It doesn't work. Then, von Guericke walks towards the hemispheres and separates the two hemispheres with his bare hands! Everyone in the crowd is in shock! In fact, the trick hinges on the small valve we mentioned earlier … and this is not an illusion, but a real, pragmatic, engineering problem. Von Guericke has actually invented one of the first pneumatic machines using piston-type vacuum pumps. Indeed, this kind of device creates a vacuum in a container by sucking the air out! Conversely, if air enters in a container previously "filled" with vacuum, the vacuum is destroyed by the incoming air. This is exactly what happens in the Magdeburg hemispheres experiment in which

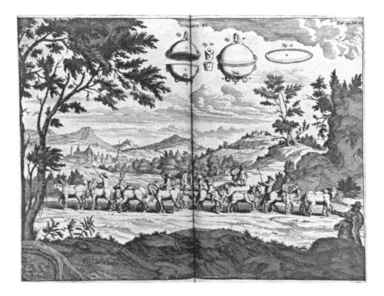

FIGURE 1.5 Magdeburg hemispheres experiment by von Guericke in "*Experimenta nova (ut vocantur) Magdeburgica de vacuo spatio*" (1672).

von Guericke plays with creating/destroying vacuum in the sphere. From a scientific standpoint, this experiment also shows the importance of atmospheric pressure. When the sphere is "full" of vacuum, there is no pressure inside the sphere; hence the pressure outside the sphere (the atmospheric pressure) is much greater than the pressure inside the sphere. It is then very difficult to separate the two hemispheres of the sphere ... but not impossible. If we calculate the forces of pressure on the sphere, we find that we need approximately 40,000 Newtons, thus the action of a mass equivalent to 5 tons! In general, a horse is only able to draw about 0.212 ton, so 16 horses were either a great guess or an excellent calculation! Let us add that vacuum technology leads to many inventions, most notably pneumatic tube transport systems. At the end of the 19th century, many large cities use this system to send and receive messages or small packages. One of the oldest is the system linking the London Stock Exchange to the city's main telegraph station. Cities such as Berlin, Paris, Geneva, Vienna, Prague, Algiers, Buenos Aires, Chicago and New York City also develop pneumatic mail services, which remain active in many cases until the 1950s. The Paris system stays operational until 1984, and the Czech pneumatic tube mail system is only stopped because of flooding in 2002, the system being damaged beyond repair. If you visit old office buildings in New York, you will certainly see remains of pneumatic mail systems. Look for the tubes!

BOOM OF EXPERIMENTAL CHEMISTRY AND FIRST EMPIRICAL LAWS

On the philosophical front, a new movement begins during the Renaissance, known as empiricism. Its birth is often attributed to Bacon (1561–1626) who describes a new way to structure and understand the sciences in his master plan "*Instauratio Magna*" (Latin: the Great Instauration). At the beginning, Bacon thinks about writing six different books, each of them describing his views on (i) how to summarize and communicate scientific results in "*De Augmentis Scientiarum*" (The Advancement of Sciences), (ii) how to understand and interpret Nature in "*Novum Organum*" (The New Instrument), (iii) how to observe natural facts in "*Historia Naturalis et Experimentalis*" (Natural and Experimental History), (iv) how to apply the inductive intellectual method in "*Scala Intellectus*" (The Ladder of the Understanding), (v) how scientific facts were established before Bacon's method in "*Prodromi Sive Anticipationes Philosophiae Secundae*" (Preconceptions or Anticipations of the Second Philosophy) and (vi) how his new philosophy will lead to the development of science by future generations in "*Philosophia Secunda aut Scientia Activae*" (The Second Philosophy or Active Science). Bacon's idea is to revolutionize the sciences with a comprehensive vision to better understand the laws of Nature. His method emphasizes especially the use of the experimental method, based on observations and trials, to establish facts. Then, to understand these facts, he submits his hypotheses to inductive reasoning, using as many comparisons and exclusions as possible. According to Bacon's new logic (contrary to Aristotle's, which is based on deduction, i.e. conclusions are drawn from postulates that are intuitively evident), knowledge can be captured through our senses, which provide information on the objects of Nature. The issue for Bacon is that we impose our own interpretations on these objects! This means that our scientific theories are constructed according to the way we see objects, therefore biasing our formulation of hypotheses. To explain this, he will develop the concept of "Idols", which are the causes of errors that we make when reasoning (note that this concept will be used again, centuries later, by Freud in psychoanalysis). There are four of them: Idols of the Tribe (reasoning mistakes common to all human beings – for instance, oversimplification or overemphasis of exciting events in our observations), Idols of the Cave (reasoning mistakes specific to individuals – for instance, overly focusing on differences between things or on details), Idols of the Marketplace (reasoning mistakes due to the inaccuracies of our language – for instance, the same word can be used to group different things together, thereby leading to fruitless discussions) and Idols of the Theater (reasoning mistakes due to fallacious argument and false learning – for instance, considering as established truth old conclusions and arbitrary divisions of knowledge). For the anecdote, let us note that Shakespeare (1564–1616) and Bacon were contemporaries. According to some, Bacon might even have contributed to the Shakespearean masterpieces.

The development by the end of the 18th century of radically novel scientific instruments, including the telescope, microscope, pendulum clock, thermometer, air pump and barometer, results in deep social changes. It can perhaps be better sensed by looking at the famous painting "An Experiment in the Air Pump" (1768) by Wright of Derby. It shows a traveling scientist who operates a vacuum pump in front of an audience at night. In addition to exhibiting this scientific device, Wright of Derby also gives us access to the broad range of human emotions prompted by this invention. Indeed, the fight between life and death of the cockatoo caused by air deprivation triggers a set of very diverse reactions, including fright for children, deep thinking for the philosopher, excitement for the young but also indifference for the young lovers. Wright of Derby adds more dramatic effect, and maybe gives us an insight into his own thoughts by using a single candle to light the scene and by displaying a glass containing a skull in the foreground of his painting.

Let us add that, following Bacon's ideas, intellectuals start to gather in a club called the Oxford (England) "Experimental Philosophy Club". It later becomes the Royal Society in the 1660s. The 12 Founder Fellows of the Royal Society (FFRS) are William Ball (astronomer), Robert Boyle (natural philosopher, inventor), William Brouncker (mathematician), Alexander Bruce (inventor, politician), Jonathan Goddard (physician), Abraham Hill (merchant), Sir Robert Moray (statesman, scientist), Sir Paul Neile (politician, astronomer), Sir William Petty (political economist), Lawrence Rooke (astronomer, mathematician), John Wilkins (clergyman, Head of Colleges both in Cambridge and Oxford) and Sir Christopher Wren (astronomer, geometrician and architect). They discuss, among other things, how to promote science. Most notably, they create, in 1665, the first periodical named the "Philosophical Transactions", which is still being published today! It marks the creation of a scientific community.

From a scientific point of view, Boyle (1627–1691) and his assistant Hooke (1635–1703) carry out a series of extremely fascinating experiments and discuss them in 1660 in "New Experiments Physico-Mechanical, Touching the Spring of the Air and its Effects" (see Figure 1.6). Among other things, they build and use an air pump to study the physical nature of air. Thanks to this device, they measure several of its properties. Their theory is that air is elastic (it returns to its original shape and size

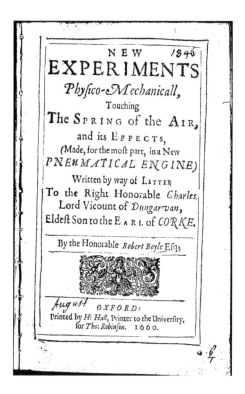

FIGURE 1.6 Cover page of Boyle's "New Experiments Physico-Mechanical, Touching The Spring of the Air and its Effects" (1660).

after being stretched or squeezed) – hence the phrase "spring of air". According to Boyle, everything can be understood through a "mechanical philosophy", in which the Universe is a gigantic machine! Here are some interesting experiments on combustion, respiration and sound transmission. Do not forget that we are just at the beginning of the 17th century. "Experiment 10" studies the burning of candles and show that the appearance of the flame changes as air pressure is gradually decreased using an air pump! "Experiment 26" shows that a pendulum swings more quickly when air is rarified! "Experiment 27" demonstrates that the ticking sound of a watch is less loud, and even vanishes, when air is removed! "Experiment 28" reveals that bubbles can suddenly form in water when air pressure is rapidly decreased! "Experiment 40" discusses the difficulty for insects to sustain flight or walk under rarefied air. Indeed, under low air pressure, insects cannot walk or fly any longer! "Experiment 41" deals with respiration and the ability of animals to breathe when modifying the air pressure surrounding them. They even succeed in reviving animals!

In a second edition published in 1662, Boyle and Hooke create a new experiment using a J-tube. The J-tube is sealed at one end and, by pouring mercury through the open end of the tube, they manage to trap a small volume of air in the sealed end of the tube. They find that the more mercury they add, the smaller the volume of the trapped air is! Their genius idea is to write down the different values taken by the volume of air and the corresponding pressure that mercury imposes on the trapped air (see

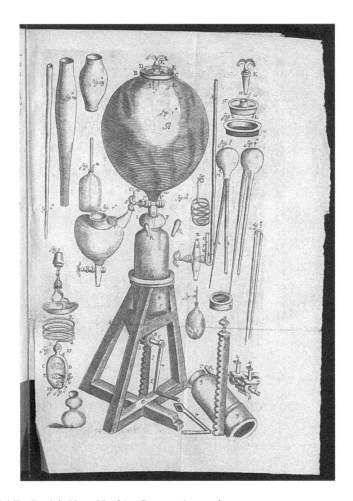

FIGURE 1.7   Boyle's *Nova Machina Pnevmatica* or air pump.

Figure 1.7). They notice something strange. For each experiment that they carry out, the product of volume times pressure always gives the same result! For instance, in a first experiment, they report a volume V_1=46 (on their graduated tube) and a pressure P_1=30 9/16 (reading for the height of mercury) leading to the product P_1V_1=1406. In a second experiment, they find V_2=16 and P_2=87 14/16, again giving a product P_2V_2=1406! And they find the same thing over and over again. They also notice something else. If they carry out a third experiment for which they pour more mercury to reach a pressure twice that of the second experiment (P_3=2P_2 = 175 12/16) they find a volume V_3=8 which is exactly 16/2=V_2/2! Similarly, when the pressure is multiplied by 3, the volume is divided by three, and so on. They finally reach the conclusion that the pressure is inversely proportional to the volume (P varies as 1/V). In other words, when the pressure exerted on the air increases, then the air volume decreases. This is Boyle's law! This law is also known under the name of Mariotte's law or Boyle–Mariotte's law. Mariotte also finds that P varies as 1/V at constant temperature in his book "Discourse on the Nature of Air" (1676), independently from Boyle.

Let us add that Boyle is also an alchemist, who, all his life, works on the secret of transmutation. Perhaps one of Boyle's most interesting books is the famous "The Sceptical Chymist" in 1661. Indeed, in this work, he postulates the corpuscularian hypothesis in opposition to Aristotle's model (four elements: water, earth, fire, air) and Paracelsus' alchemy (three basic principles: mercury, salt, sulfur). From his numerous experiments, he claims that air is composed of corpuscles that move, with vacuum between them. This is the reason why it is possible to compress or expand air. Boyle even thinks that these corpuscles can associate and form clusters during collision.

Proposition I: "It seems not absurd to conceive that at the first Production of mixt Bodies, the Universal Matter whereof they among other Parts of the Universe consisted, was actually divided into little Particles of several sizes and shapes variously mov'd."

Proposition II: "Neither is it impossible that of these minute Particles divers of the smallest and neighbouring ones were here and there associated into minute Masses or Clusters, and did by their Coalitions constitute great store of such little primary Concretions or Masses as were not easily dissipable into such Particles as compos'd them."

FIRST BALLOON FLIGHTS AND GAS LAWS

We owe another important discovery to Charles (1746–1823). Thanks to him and others, the history of aeronautics begins! Charles' brilliant idea is to use Cavendish's finding, during the second half of the 17th century, that hydrogen is less dense than air. Indeed, on August 27, 1783, he makes the first balloon filled with hydrogen (and names it a "*Charlière*") fly above the Champ de Mars in Paris (see Figure 1.8), where the Eiffel Tower is today. This is an incredible feat! Note that a certain Benjamin Franklin is in the crowd.

The first flight with living creatures takes place on September 19, 1783, with a sheep, a duck and a rooster placed in a basket attached to a balloon. This happens in Versailles in front of King Louis XVI and Queen Marie Antoinette. However, this time, this is a hot-air balloon, created by the Montgolfier brothers, Joseph-Michel (1740–1810) and Jacques-Etienne (1745–1799) (see Figure 1.9). It operates according to the following principle. Let us recall that Amontons (1663–1705), a hundred years earlier, develops the air-pressure thermometer and shows that air pressure increases with temperature. As the balloon is subject to its own weight and Archimedes' upward buoyant force, the balloon can only rise (or fly) if the buoyant force is greater than the weight of the balloon! Since Archimedes' buoyant force is equal to the weight of the volume of displaced air, the upward force will be greater when the volume of the balloon is larger. In fact, following Amontons' law, if one uses air hotter than the outside air, the pressure increases inside the balloon. It then allows the balloon to expand (the pressure inside pushes back the envelope, increasing its volume) thus increasing the upward force, making the balloon go up. It is therefore possible to make a balloon go up (or down) by heating (or cooling) the air inside the balloon! For the anecdote, Charles and Robert will fly their first manned flight balloon filled with hydrogen, ten days after the first manned "*Montgolfière*" flight! Maybe this

FIGURE 1.8   First hydrogen balloon flight in Paris (1783).

is the reason why Charles will do many experiments on gases later on, which will eventually lead to Charles' famous law.

Gay-Lussac (1778–1850) does himself a significant amount of ballooning. In fact, he sets the record for the highest balloon flight at 7,016 meters (about 23,000 feet), a record that holds for more than 50 years. This is maybe why he is so interested in Charles' research. Even if Charles does not publish his results, Gay-Lussac builds on them and adds results from his own experiments using air thermometers. For this purpose, he measures the air volume trapped in a tube by pouring mercury. His idea is then to immerse the tube in water at different temperatures. He realizes that, to maintain the same volume of air, he needs to either add or remove mercury depending on the temperature. In particular, he observes that the volume (V) increases when the temperature (T) increases! Moreover, when he looks at the results from two different experiments, carried out at two different temperatures, T_1 and T_2, he

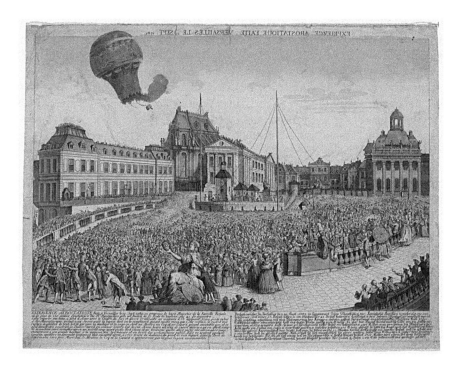

FIGURE 1.9   Balloon launched by the Montgolfier brothers in Versailles (1783).

observes that $T_1V_2 = T_2V_1$! He finds the same formula for different gases such as oxygen, nitrogen and hydrogen. He names this relation Charles' law in Charles' honor. He also notes that, if he draws a graph for the volume against the temperature, a volume of gas of zero ($V = 0$) is reached for a temperature of $-266.66°C$! For Gay-Lussac, this shows that Charles's law must be taken with caution and does not apply at low temperatures. Nevertheless, Lord Kelvin will give another meaning to this number representing the "infinite cold" or absolute zero, leading to an entirely new science, now known as thermodynamics!

Gay-Lussac also uses the data for the different heights of mercury from his experiments. He deduces that the pressure of the gas increases when temperature increases. Repeating the same process as before, i.e. looking at two experiments for two different temperatures, he finds that $P_1T_2 = P_2T_1$! This law is called Gay-Lussac's law. Gathering Boyle's, Charles' and Gay-Lussac's laws, we obtain the famous combined gas law that gives the relationship between temperature, pressure and volume for any gas. Using mathematical equations, this translates into $PV = constant$ (Boyle's law), $V/T = constant$ (Charles' law) and $P/T = constant$ (Gay-Lussac's law). It thus follows that $(PV)/T = constant...$ Q.E.D.! This is a very powerful formula that can be used to solve practical problems. For instance, consider a gas at temperature T_1, with volume V_1 and pressure P_1. If we increase the temperature to twice the initial value ($T_2 = 2T_1$) and keep the same volume ($V_2 = V_1$), we can calculate the pressure $P_2 = (P_1V_1T_2)/(V_2T_1)$. We thus find that, since $V_2 = V_1$ and $T_2 = 2T_1$, $P_2 = 2P_1$! In other words, when a gas is enclosed in a vessel of fixed volume and we warm it up to twice the initial temperature, we find that the pressure at this new temperature is twice the value it had at the beginning of the experiment! We can also solve much more complex problems, for which two variables change, for instance, P and V, or P and T, or T and V. We just need to use the fact that $(P_1V_1)/T_1 = (P_2V_2)/T_2$ for any gas! This kind of calculation is very important, as it will be an essential tool in the advent of steam engines!

Let us mention another important empirical law, named after Dalton (1766–1844), and sometimes called the law of partial pressures. As mentioned in his paper "On the Absorption of Gases by Water and other Liquids", read in front of the Literary and Philosophical Society of Manchester in 1803, he presents his discovery that air is a mixture of different gases! He finds that air is mainly composed of nitrogen (N_2), oxygen (O_2), carbon dioxide (CO_2) and water vapor (H_2O). Moreover, he determines that the pressure for air is equal to the sum of the pressures for each of the gases, taken separately. Expressed differently, each gas (labeled "i") contributes its partial pressure (P_i), yielding the total pressure of air as $P_{air} = P_{O_2} + P_{N_2} + P_{CO_2} + P_{H_2O}$ or, in shorthand, $P_{air} = \sum P_i$. Let us add that, for simplification, many authors often define air as composed of 80% of N_2 and 20% of O_2, the other gases being only present as traces. To better understand the reach of Dalton's finding, consider the following experiment. Let us take a box with two compartments separated by a wall, the first compartment containing 1 liter of N_2 at a pressure of 1 bar and the other holding 1 liter of O_2 also at 1 bar. At the start, the two compartments are separated by the wall and no gas flow takes place. Removing the wall and allowing for the two gases to mix, we find that the total pressure in the entire box does not vary; it is of 1 bar, even if the total volume is now 2 liters! When we put back in place the wall between the two compartments, we find that each compartment of 1 liter, taken separately, is composed of a gas mixture with 50% of N_2 and 50% of O_2 and is at a pressure of 1 bar. This example illustrates two important points. First, at a given temperature, the pressure of each gas in the gas mixture is equal to the pressure of this gas taken separately in the total volume. In our example, at the start of the experiment, when N_2 occupies a volume of 1 liter, $P_{N_2} = 1 bar$ (and the same for O_2). After mixing, we end up with a volume of 2 liters, hence, according to Boyle's law, the pressure of N_2 is now half of what it was initially ($P_{N_2} = 0.5$ bar). Similarly, $P_{O_2} = 0.5$ bar. Second, at a given temperature, the total pressure of a gas mixture is equal to the sum of the partial pressures of the gases composing it. Again, in our example, we can see that $P_{tot} = P_{N_2} + P_{O_2}$, with the partial pressures being $P_{N_2} = 0.5$ bar and $P_{O_2} = 0.5$ bar. Practically, this has very significant consequences for activities such as space travel or, closer to us, scuba diving. Indeed, astronauts and divers both need air tanks to breathe. For instance, when diving to a depth of about 20 meters (66 feet) below sea level, the air pressure is about 2 bar, meaning that the partial pressures are $P_{O_2} = 0.4$ bar (20% $\times$ 2 bar) and $P_{N_2} = 1.6$ bar (80% $\times$ 2 bar). These numbers are very different from the partial pressure at sea level, with $P_{O_2} = 0.2$ bar (20% $\times$ 1 bar) and $P_{N_2} = 0.8$ bar (80% $\times$ 1 bar). From a biochemistry standpoint, diving accidents occur when some gases become toxic to the human body. For example, if the partial pressure $P_{O_2} > 1.6$ bar (beyond 70 meters or 230 feet), there is a risk of hyperoxia (when cells, lungs and body tissues are exposed to an excess of oxygen). For $P_{N_2} > 3.2$ bar (beyond 35 meters or 115 feet) there is a risk of nitrogen narcosis (when having nitrogen at high pressure in the body causes an anesthetic effect). Let us note, on a theoretical level, Dalton's law is also very significant because it implies that each gas acts as if it does not feel the presence of other gases in a mixture. This observation will be a building block of Dalton's atomic theory formulated in 1808 in "A New System of Chemical Philosophy".

The last important law is Avogadro's law. It stipulates that, at the same temperature and pressure, two gases enclosed in equal volumes will have the same number of molecules, regardless of the nature of the gas studied! Put differently, the number of molecules in this volume, at this pressure and temperature, will always be the same for any gas studied! This is a huge statement that has a lot of implications. To better understand its considerable significance, let us have a look at Gay-Lussac's law of combining volumes. Indeed, from his experiments, Gay-Lussac discovers that 2 volumes of hydrogen gas and 1 volume of oxygen gas are needed to obtain 2 volumes of water vapor at a given temperature and pressure. Using a mathematical equation, it gives $2 V_{H_2} + 1 V_{O_2} \rightarrow 2 V_{H_2O}$. This means that, to have the same volume, and thus the same number of molecules, the volume of hydrogen, as well as the volume of water vapor, should be divided by 2. This leads to $V_{H_2}/2 = V_{O_2}/1 = V_{H_2O}/2$ or, in shorthand, V/n = constant which is called Avogadro's law!

By combining all the laws presented above, Clapeyron (1799–1864) in his memoir of 1834 proposes the following law: $Pv = R(267 + t)$, where t is in degrees centigrade (the number 267 is to be connected to the value of –266.66 found by Gay-Lussac). Indeed, he finds that the ratio $Pv/(267 + t)$ is a constant. It seems that the letter "R" comes from the French word "rapport" (ratio in English). Later, in 1864, Clausius (1822–1888) gives a simplified version of this equation, the famous $Pv = RT$ which is close to the form we use today. Let's note that, in both equations, the volume is taken per unit mass of gas (v = V/M) and R varies from one gas to another. It is finally Horstmann (Clausius' student) who, in 1873, gives us the more modern equation (uP = RT), in which he uses the notion of "mole" through the volume per mole u (u = V/n) and finds that R is a universal constant, i.e. the same for all gases! Nowadays, we use the following equation for an ideal gas: $PV = nRT$ with $R = 8.314$ Pa m^3 K^{-1} mol^{-1} or 8.314 J K^{-1} mol^{-1}, meaning that there is a relation between J (for Joule) and Pa m^3. Or in other words, energy (J) is related to the product PV (Pa m^3), also known as work, defined by a certain James Prescott Joule.

RATIONALISM, MODERN MATHEMATICS AND MOTION

Another great thought current emerges during the Scientific Revolution: rationalism. As the name suggests, it is based on the use of reason, otherwise known, at the time, as common sense. It starts with Aristotle's concept of λόγος (*logos*) (Greek = speech, reason), that distinguishes humans from animals. The use of reasoning becomes a philosophical principle with Descartes. According to him, the ability to "think well" is present in all men. The only difficulty in achieving it is to avoid departing from the righteousness (or logic) of our thoughts because of passions, appetites, desires, opinions or preconceptions. For him, "the diversity of our opinions does not come from the fact that some are more reasonable than others, but only that we conduct our thoughts in various ways, and do not consider the same things". In the "Discourse on the Method" (1637), he shares his experience and explains the method he developed to lead his life. This is intended as a starting point for us to think and find our own method for our own life. The use of logic as a tool for reasoning appears even more clearly in Spinoza's writings. For instance, his main work, "Ethics" (1677), consists of definitions, propositions, axioms, proofs, corollaries, lemmas ... like a mathematics book with no equation! It deals with the existence of God, the nature and origin of the mind, human emotions, as well as the power of human understanding and human freedom. For Leibniz (1646–1716), who is both a mathematician and a philosopher, logic is at the center of everything. He proposes in *"Monadologia"* (1714), a metaphysical treatise that summarizes his system of thought, regrouping the monads (the elements of the world, their nature and degrees of perfection), God (the creator of the world, his existence and nature) and the created world (the world itself and its universal harmony), with the two principles founding his philosophy: the principle of contradiction and the principle of sufficient reason. "Our reasonings are based on two great principles, that of contradiction, in virtue of which we judge that which involves a contradiction to be false, and that which is opposed or contradictory to the false to be true. And that of sufficient reason, by virtue of which we consider that we can find no true or existent fact, no true assertion, without there being a sufficient reason why it is thus and not otherwise, although most of the time these reasons cannot be known to us." Finally, let us add that Kant (1724–1804) will push these concepts of rationalism and empiricism even further by founding transcendental idealism. It is often described by the difference between *"das Ding an sich"* (the thing-in-itself) and *"das Ding für mich"* (the thing-for-me). The way we perceive the world through our sensations is just as important for Kant as it is for empiricists. Nevertheless, he introduces the fact that the way we perceive the world around us is inherent to our way of looking at it, of analyzing it through our reason. "I call all knowledge transcendental if it is occupied, not with objects, but with the way that we can possibly know objects even before we experience them." In particular, he studies the possibility of a scientific knowledge through the theoretical use of reason in "Critique of Pure Reason", morality through the practical use of reason in "Critique of Practical Reason" and the aesthetic in "Critique of Judgment".

The development of rationalism is accompanied by mathematical advances. From an historical perspective, one of the first fundamental books in mathematics is certainly Euclid's "Elements" (mid-4th century BC–mid-3rd century BC). It gathers the essential knowledge of geometry accumulated during Greek antiquity. We find different concepts, including plane geometry with famous theorems such as the Pythagorean and Thales' theorems and number theory with prime $(2,3,5,7,11,13,17, \ldots)$, rational $(1/2, 2/5, 6/7, \ldots)$ and irrational numbers $(\pi, \sqrt{2}, \ldots)$, as well as the geometry of solids with the famous five Platonic solids.

It is during the 17th century that Descartes proposes an analytic geometry which aims at combining geometry and algebra (see Figure 1.10). He thinks of representing an object with a mathematical equation. For example, a point is characterized by its "Cartesian" (from Descartes) coordinates in a Euclidean (from Euclid) plane. A line becomes an equation, a circle another equation. Thanks to this representation, we can also calculate a distance between two points, which we call today the Euclidean metric. Note that there are other metrics, or ways of measuring a distance, such as the use of angles in elliptic or hyperbolic geometries.

However, one can be both empiricist and rationalist. The most famous example is certainly Newton (1642–1727). Through his work in geometry and physics, Newton goes even further than Descartes and defines a mathematical formalism to describe the motion of a physical object! In other words, he introduces the notion of time in equations! This revolution begins with his work in *"De methodis serierum et fluxionum"* (On the Methods of Series and Fluxions) completed in 1671. Studying

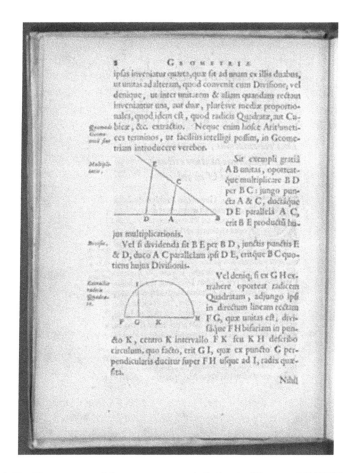

FIGURE 1.10   Descartes combines geometry and algebra in *"Geometria"* (1664).

quantities that vary with time, he has the genius idea of inventing the famous concepts of "fluents" and "fluxions". In the English translation from 1737, we can read: "Now thofe quantities which I confider as gradually and indefinitely increafing, I fhall hereafter call Fluents, or flowing Quantities, and fhall reprefent them by the final letters of the alphabet v, x, y and z; that I may diftinguifh them from other quantities, which in equations may be confidered as known and determinate, and which therefore are reprefented by the initial letters a, b, c. And the velocities by which every Fluent is increafed by its generating motion (which I may call Fluxions, or fimply Velocities, or Celerities,) I fhall reprefent by the fame letters pointed thus, $\dot{v}$, $\dot{x}$, $\dot{y}$ and $\dot{z}$; that is, for the celerity of the quantity v I fhall put $\dot{v}$ and fo for the celerities of the other Quantities x, y, and z, I fhall put $\dot{x}$, $\dot{y}$ and $\dot{z}$, refpectively." Then he ends up with two fundamental problems. (1) Calculate the velocities, given the quantities, and conversely. (2) Calculate the quantities, given the velocities. To solve these two problems, he develops the concept of "Moments of flowing quantities" as "their indefinitely fmall parts, by the acceffion of which, in indefinitely fmall portions of time they are continually increas'd". It allows him to calculate the value of a quantity after an infinitesimal (or indefinitely small) interval of time. For instance, if the quantity is x, its new value after an infinitesimal time interval is $x + \dot{x}o$ ($\dot{x}o$ being the product of the velocity of x times an "indefinitely fmall quantity o" of time). The same reasoning applies to the other quantities v, y and z. Then, he proposes to take the "ultimate ratio" defined as $(x(t_2) - x(t_1))/(t_2 - t_1)$, in which $x(t_2) = x + \dot{x}o$, $x(t_1) = x$ and $t_2 - t_1 = o$. Plugging this into the equation for the "ultimate ratio", we find $(x + \dot{x}o - x)/(o) = \dot{x}$, the famous fluxion of the quantity x! Newton justifies the "ultimate ratio" as characteristic of motion. "For by the ultimate velocity is meant that with which the body is moved, neither before it arrives at its last place and the motion ceases, nor after but at the very instant it arrives [...] And in like manner, by the ultimate ratio of evanescent quantities is to be understood the ratio of the quantities, not before they vanish, nor afterwards, but with which they vanish." Mathematically, if we have the equation (or fluent)

$$x^2 + y^2 - 4 = 0 \tag{1.1}$$

and substitute x with $\dot{x}o$ and y with $\dot{y}o$, we obtain

$$(x + \dot{x}o)^2 + (y + \dot{y}o)^2 - 4 = 0$$

This leads to

$$x^2 + 2x\dot{x}o + (\dot{x}o)^2 + y^2 + 2y\dot{y}o + (\dot{y}o)^2 - 4 = 0 \tag{1.2}$$

By calculating Equation 1.2 minus Equation 1.1, we have

$$2x\dot{x}o + (\dot{x}o)^2 + 2y\dot{y}o + (\dot{y}o)^2 = 0$$

We then divide by "o" to obtain the ultimate ratio and find

$$2x\dot{x} + \dot{x}^2 o + 2y\dot{y} + \dot{y}^2 o = 0$$

Following Newton's method, we "reject" the terms that are multiplied by "o" as it "will be nothing in refpect of the reft". This results in $2x\dot{x} + 2y\dot{y} = 0$, also known as the fluxion of the fluent $x^2 + y^2 - 4 = 0$! The first fundamental problem is therefore solved! Let us note that Equation 1.1 is an equation describing a circle of radius 2, and that we've just traveled on that circle for an infinitesimal amount of time! In order to solve the second fundamental problem previously mentioned, Newton manages to calculate the fluxions of many fluents, which he gathers in tables. Doing this, he relates fluxions to fluents, allowing him to identify the fluent associated to a specific fluxion when needed. Nowadays, we use a modern version of these concepts in the differentiation and integration formulas found in mathematics books.

Leibniz (1646–1716) also works on the famous "problem of tangents" and develops the calculus of the "infinitely small" or infinitesimal. Like Newton, he also discovers the "backward differentiation", sometimes called the "inverse problem of tangents", through integral calculus. Both compose what is today the body of knowledge covered by calculus in mathematics! Leibniz first presents his findings in "*Nova Methodus pro Maximis et Minimis, Itemque Tangentibus, qua nec Fractas nec Irrationales Quantitates Moratur, et Singulare pro illi Calculi Genus*" (A New Method for Maxima and Minima as well as Tangents, which is Impeded Neither by Fractional nor by Irrational Quantities, and a Remarkable Type of Calculus for This) (1684). In particular, he defines the following rules for differentiation: $d(x+y) = dx + dy$ and $d(xy) = xdy + ydx$. Let us consider the same example as before, $x^2 + y^2 - 4 = 0$ (1.1) and examine it according to Leibniz's "tangent problem". To do this, we start from a point (x,y) on the circle and move of an infinitesimal distance dx along the x axis and of dy along the y axis to a new point (x+dx, y+dy). Then, following Leibniz, we can find the slope of the tangent as the quotient of these two infinitesimal distances dy/dx. So, first, let us apply Equation 1.1 at the new point (x+dx,y+dy), which gives

$$(x+dx)^2 + (y+dy)^2 - 4 = 0$$

Expanding this, we obtain

$$x^2 + 2xdx + dx^2 + y^2 + 2ydy + dy^2 - 4 = 0$$

Subtracting Equation 1.1 to this new equation gives

$$2xdx + dx^2 + 2ydy + dy^2 = 0$$

As suggested by Leibniz, let us neglect the squared terms dx^2 and dy^2 to obtain $2xdx+2ydy=0$. After dividing by dx, this gives $2x+2y(dy/dx)=0$. Rearranging the equation, the fractional form emerges $(dy/dx)=-x/y$, which is the slope of the tangent! Let us add that it is only during the 19th century that the concept of limits appears, giving a better explanation for the "disappearance" of the squared terms dx^2 and dy^2, and finally providing a correct definition of derivatives. Two years later, in "On a Deeply Hidden Geometry", Leibniz discovers the concept of antiderivatives, like Newton, which propels calculus into a new era and forms the basis of the fundamental theorem of calculus! If we look at our previous example, we know that $dy/dx=-x/y$ which leads to $ydy=-xdx$. Integrating both sides, we obtain $\int ydy=-\int xdx$. We then have $y^2=-x^2+$const. or y^2+x^2-const. $= 0$. We therefore recover Equation 1.1 by identifying const. $= 4$ and finally find $y^2+x^2-4=0$! Let us add that one of the most remarkable discoveries of Leibniz is his famous formula $\pi/4=1-1/3+1/5-1/7+\ldots$, obtained by calculating the area of a quarter-circle of radius 1 using integral calculus! Moreover, Leibniz also introduces many mathematical symbols that are currently used such as the dot (.) for multiplication, the colon (:) for division, the integral (the long S for sum $\int$) and the infinitesimal increments on the abscissae (dx) and ordinates (dy), among others.

Now let us go back to Newton. Applying his mathematical discoveries to the physical phenomena surrounding him, Newton publishes in 1687 his *"Philosophiae Naturalis Principia Mathematica"* (Mathematical Principles of Natural Philosophy). In this book, he defines the fundamental principles of mechanics (see Figure 1.11). By mechanics, Newton means that the world around us is akin to a machine whose operation can be understood and predicted using mathematics. His famous laws of motion are (i) the principle of inertia (in the absence of external forces, an object follows a linear trajectory at constant speed), (ii) the fundamental principle of dynamics (the product of the mass m of an object by its acceleration **a** is equal to the sum of the external forces **F** exerted on the object: m**a** = $\sum$**F**), (iii) the principle of action–reaction (for two interacting objects, there is a pair of forces, of equal magnitude and opposite directions, acting on the two objects).

From these three laws, Newton formulates his well-known law of gravitation, finally providing an explanation of the empirical laws of Kepler. They describe the main features of the movement of the planets around the Sun, thereby demonstrating the validity of heliocentrism and Copernican theory! On Earth, gravitation manifests itself as the terrestrial attraction, or gravity, that keeps us on the ground, and makes apples fall! Now, considering the case of gases, such as, for instance, air enclosed in a container, gravity should lead to the aggregation of its molecules at the bottom of the container. However, we see in experiments that no such clusters form. It seems that something else is happening!

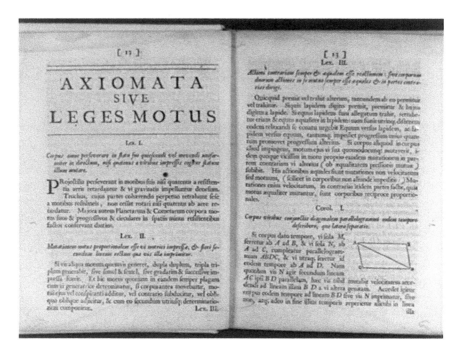

FIGURE 1.11 Newton's laws of motion in *"Philosophiae naturalis principia mathematica"* (1687).

MOTION TO ACTION: THE BIRTH OF ANALYTICAL MECHANICS

And this something has to do with the famous momentum ($\mathbf{p} = m\mathbf{v}$) of an object which represents the concept of "mass in motion", and, in particular, how it changes when this object comes into contact with another object. Descartes is the first to introduce the rules and the laws that govern this extremely fast phenomenon called collision between particles. In fact, in *"Principia philosophiae"* (1644), Descartes writes that (i) "The first law of nature: that any object, in and of itself, always perseveres in the same state; and thus what is moved once always continues to be moved", (ii) "the second law of nature is any part of matter, considered apart, never tends to continue to be moved along any oblique lines, but only along straight lines, even if many are often forced to deflect due to the collision of others", (iii) "Third law: that a body, in colliding with another larger one, loses nothing of its motion; but, in colliding with a smaller one, loses as much as it transfers to that one." He notices that, at the time of the collision between two bodies, something seems to be transferred to the other body. Descartes' intuition together with Newton's law of motion shows that during the collision the total change in momentum is zero or that there is conservation of the overall momentum!

Shortly after, Huygens (see Figure 1.12) resumes Descartes' experiments on collisions, that, according to him, are central to our understanding of the development of the Universe! Specifically, he finds that the momenta (Latin: plural for momentum) before the collision (initially: $\mathbf{p}_i = m_i \mathbf{v}_i$) and after the collision (finally: $\mathbf{p}_f = m_f \mathbf{v}_f$) are conserved ($\mathbf{p}_i = \mathbf{p}_f$), in line with Newton and Descartes' ideas. He also notes that

FIGURE 1.12   Christiaan Huygens (1629–1695).

the quantities (or scalars) mv^2, before and after the collision ($m_i v_i^2$ and $m_f v_f^2$), are also conserved! He publishes his results in 1669. His student Leibniz continues his work on collisions and derives the famous law of conservation of kinetic energy, as it will be called after. For this purpose, he starts from the idea that there is a form of energy connected with motion. This famous quantity, mv^2, is a number, and not a vector as it is for momentum. He names it "*vis viva*" (Latin: living force) of the body. He publishes his results in 1686 but receives negative criticisms. At the time, many scientists choose to support Newton and Descartes' arguments, given their notoriety. Note that, much later, it will be demonstrated that both quantities (momentum and kinetic energy) are indeed conserved during an elastic collision.

   The law of conservation of *vis viva* will be very useful to Bernoulli (1700–1782). Indeed, in "Hydrodynamica" (1738), he notices that the loss of *vis viva*, when water flows, is related to the change in pressure, the famous Bernoulli's principle! This principle is very important since this is the first time that the motion of a fluid (as a smooth and continuous collection of molecules) can be understood and analyzed, a task impossible with Newton's formalism. This is the birth of fluid mechanics! Bernoulli will also write the first kinetic theory of gases based on the conservation of *vis viva*! To achieve this, he uses a device comprised of a movable piston in a cylinder filled with gas. According to him, since a gas is composed of many molecules moving in all directions, their impacts on the surface of the cylinder give rise to pressure, and their persistent motion results in the heat he feels when touching the cylinder. He also notes that the intensity of the *vis viva* is proportional to the temperature of the gas!

This notion of *vis viva* is also at the origin of what is now called analytical mechanics. Indeed, this branch of mechanics is dedicated to the determination of the trajectory of an object. Its advent is often attributed to Maupertuis (1698–1759) for his principle according to which the trajectory followed by an object will have the shortest length: "Nature always acts in ways that are the simplest and the shortest". To do this, he applies the concept of integral (which mathematically stands for the path) to the equations of motion of the object given by Newton's laws. Another version of what is commonly called the principle of least action is Fermat's principle, which states that a ray of light follows the path of shortest duration. It is finally Euler (1707–1783) who, in 1744, gives the mathematical formulation of the principle of least action in "*Methodus inveniendi lineas curvas maximi minimive proprietate gaudentes*" (A Method for Finding Curved Lines Enjoying Properties of Maximum or Minimum). Indeed, he follows Maupertuis' idea: "When a change occurs in Nature, the amount of action necessary for this change is the smallest possible. The amount of action is the product of the mass of bodies by their velocity and by the distance they travel. When a body travels from one place to another, the action is greater if the mass is greater, the velocity higher and the distance traveled further." Thanks to his formalism based on Maupertuis' principle, Euler succeeds in finding the solution to the "two-body Kepler problem" (motion of a planet about a gravitational center – for instance, the Earth about the Sun – see Figure 1.13) by solving mathematical equations! In 1772, following their work on the "three-body problem", Lagrange and Euler receive jointly the prize from the Académie des Sciences. A few years later, Lagrange (1736–1813) gives a formal proof for Maupertuis' principle and provides a general mathematical framework, and more specifically, the famous Euler–Lagrange equation! His work is published in "*Mécanique Analytique*" in 1788, which is the starting point for the new field of analytical mechanics. Thanks to this new framework, the mystery of the "three-body" system (Moon–Earth–Sun) is solved, and the motion of the Moon is explained!

Around 1830, Hamilton (1805–1865) proposes a new theory that unifies mechanics and optics. Like Lagrange, his theory is based on a variational principle (integral minimization in mathematics which corresponds to finding the shortest path in mechanics). With Jacobi (1804–1851), they develop a new formulation now known as Hamiltonian mechanics. This new mechanics revolves around the concept of energy. Later on, the increasing importance of mathematics will lead to the concept of operators to study position, momentum and energy. One of the most significant operators is undoubtedly the energy operator in quantum mechanics, that we all know through the famous equation $\hat{H}\psi = E\psi$, where $\hat{H}$ (energy operator) is called the Hamiltonian!

GAMES OF CHANCE AND MAXWELL– BOLTZMANN STATISTICAL MECHANICS

The 17th century also witnesses the birth of statistical mechanics. It is based, as its name indicates, on mechanics, i.e. the laws of motion followed by objects, combined

FIGURE 1.13 Kepler's first two laws of planetary motion (1609).

with probability theory built on statistical distributions. The latter appears with the advent of games of chance such as rolling dice, flipping a coin, playing cards. Since these games are sometimes related to money, it very quickly becomes interesting to invent a new formalism around the concept of probability with the aim of predicting a gain or a loss. At the center of this enterprise, we find Pascal, Huygens, Leibniz and many others.

Let us start from the beginning and the Bernoulli distribution (see Figure 1.14). For example, if someone asks a "yes–no" question, what is the probability of getting a "yes" answer? It is 50%, the same as getting a "no" answer, provided that there is no bias here. Now, if we look at a coin toss, we also have a 50% chance that the "heads" side turns up and a 50% chance for the "tails" side to show. Again, this is only true if there is no bias (the coin is not rigged). On the other hand, if the coin is rigged in favor of the "tails" side, the probability of drawing "tails" will be greater than of drawing "heads". Mathematically, this translates into having a probability

JACOBI BERNOULLI,

Profeſſ. Baſil. & utriuſque Societ. Reg. Scientiar.
Gall. & Pruſſ. Sodal.
MATHEMATICI CELEBERRIMI,

ARS CONJECTANDI,

OPUS POSTHUMUM.

Accedit

TRACTATUS

DE SERIEBUS INFINITIS,

Et EPISTOLA Gallicè ſcripta

DE LUDO PILÆ
RETICULARIS.

BASILEÆ,
Impenſis THURNISIORUM, Fratrum.

cIɔ Iɔcc xiii.

FIGURE 1.14   Bernoulli's *"Ars Conjectandi"* (1713).

p for "tails" such that $p > 50\%$ or $p > 0.5$. For instance, let us choose a coin for which $p = 0.6$. The Bernoulli distribution tells us that the probability q of drawing "heads" will be $q = 1 - p$, meaning that $q = 0.4$. In conclusion, with this rigged coin, we have a 60% chance for a "tails" outcome and a 40% chance for "heads"! In computer science, bits (0 or 1) are the equivalent of "yes–no" questions, laying the groundwork for machine-learning algorithms!

Another well-known distribution is the famous binomial distribution. It is based on performing several Bernoulli trials (e.g. "yes–no" questions) successively. Indeed, the idea is to calculate the probability for which a sequence of n independent experiments, or trials, will have a specific number k of successful outcomes. As an aside, the Bernoulli distribution can be thought of as a binomial distribution with $n = 1$. If the probability of success ("yes" answer) is p and the probability of a "no" answer is $(1 - p)$, then having k successes will occur with probability p^k. By the same token, getting $(n - k)$ "no" answers will occur with probability $(1 - p)^{n-k}$. The key now is to keep in mind that the k successes can take place anywhere during the n experiments, so there are many possible ways of actually getting k successful outcomes. That number is in fact represented as $\binom{n}{k}$ and is equal to the ratio $n!/(k!(n - k)!)$. The value of this number can be easily found by using the famous Pascal's triangle (the same Pascal who worked on pressure!), without performing the calculation of this ratio! In summary, the probability of a binomial distribution is written as $\binom{n}{k} p^k (1-p)^{n-k}$. If we go back to the example of the rigged coin ("heads" $p=0.6$, "tails" $q=1-p=0.4$) and flip the coin four times in a row, we find the probability of

1) Getting zero "heads": $\binom{4}{0} p^0 (1-p)^{4-0} = 0.0256$ or 2.56%
2) Getting one "heads": $\binom{4}{1} p^1 (1-p)^{4-1} = 0.1536$ or 15.36%
3) Getting two "heads": $\binom{4}{2} p^2 (1-p)^{4-2} = 0.3456$ or 34.56%
4) Getting three "heads": $\binom{4}{3} p^3 (1-p)^{4-3} = 0.3456$ or 34.56%
5) Getting four "heads": $\binom{4}{4} p^4 (1-p)^{4-4} = 0.1296$ or 12.96%

Nowadays, it has a huge impact in cryptography as some cipher algorithms are based on a binomial distribution, allowing data encryption and decryption.

But the story of distributions does not stop there. After reading Clausius' work on the diffusion of molecules, Maxwell (1831–1879) (see Figure 1.15) decides to develop a velocity distribution for particles of a gas. Indeed, he thinks that molecules move at different speeds as time goes by. He has the idea of replacing the average speed, used in previous kinetic gas theories, by a distribution of these speeds! One of the breakthroughs allowed by the use of this distribution is the estimation of the mean free path (distance traveled by a particle between two collisions). This, combined with experimental results on diffusion and viscosity, validates definitely the hypothesis of collisions between gas particles! Thanks to these calculations, Loschmidt (1821–1895) calculates the first estimate for the size of a molecule in 1865! He finds that an air molecule has a dimension of about 10^{-7} cm (or 1 nanometer, twice the value that we know now). Finally, Boltzmann (1844–1906) decides to rewrite the distribution of Maxwell, using the laws of mechanics, and formulates the famous Maxwell–Boltzmann distribution.

FIGURE 1.15   James Clerk Maxwell (1831–1879).

Mathematically, the Maxwell–Boltzmann distribution corresponds to a *chi* distribution with three degrees of freedom, which are the three spatial components (along x,y,z) for the velocity, with a scaling parameter proportional to $\sqrt{(T/m)}$ (square root of temperature over the particle mass).

We can therefore access a lot of information. For example, we can determine the most probable velocity of a molecule without carrying out any experiment! Indeed, it corresponds to the mathematical equation $df/dv = 0$, or simply to the maximum value of the distribution. Similarly, we can calculate the average velocity $<v> = \int vf(v)dv$ which also corresponds to the expected value of the velocity distribution. We can also find the root mean square velocity, $\sqrt{<v^2>} = \sqrt{(\int v^2 f(v)dv)}$, or we can bypass the integration and use the fact that variance $= <v^2> - <v>^2$ which leads to $<v^2> = $ variance $+ <v>^2$ and then take the square root of both sides. This gives the root mean square velocity which corresponds to the speed of a particle with median kinetic energy. This will lead later to the famous equipartition principle.

Let us add that Boltzmann states his famous H-theorem (Eta-theorem) in 1872. It states that a certain quantity H (Eta – Greek: lowercase: η; uppercase: H) has a tendency to decrease. Indeed, by plugging the Maxwell–Boltzmann distribution in his famous Boltzmann equation dealing with collisions, he notes that the function $H = \int f \log f \, dv$ reaches a minimum! On the contrary, if he uses another distribution, H takes a larger value, but we will come back to this point later when we discuss the second law of thermodynamics. Let us note that in 1877, Boltzmann introduces a new discipline nowadays called statistical thermodynamics, in which he relates entropy (thermodynamics) and statistics (probabilities) with the famous equation $S = k_B \ln W$!

AGITATION AND COLLISIONS: THE KINETIC THEORY OF GASES

The kinetic theory of gases aims at calculating macroscopic properties such as pressure, temperature, volume and viscosity without performing any experiment! Indeed, as its name indicates, this is a theory, and, as such, it is based on mathematical equations. Here, it builds on the mechanical laws relying on the microscopic features of the gas, such as mass, velocity and momentum of the particles that make up the gas and are invisible to the naked eye. It's a real tour de force! Let us take a closer look at this theory. As we have seen before, it starts with the work of Bernoulli, and continues with Waterston and Krönig. With the introduction of the mean free path (distance traveled between collisions) in 1859 by Clausius, everything changes! In fact, Clausius demonstrates that all molecules in a gas do not move at the same speed. And building on this observation, Maxwell makes the first step towards a velocity distribution for the molecules! The mathematical formulation of Maxwell and Boltzmann is the one we now use, as it gives a complete picture of the statistical and mechanical behavior of the molecules in a gas. However, as with any theory, it starts with hypotheses, but has limitations.

Let us begin with the hypotheses.

H1: A gas is composed of molecules in ceaseless random motion due to thermal agitation.

H2: The molecules' velocities are drawn from the Maxwell–Boltzmann velocity distribution.

H3: The molecules collide (elastic collisions) with the walls of the container in
which they are confined.

The same hypotheses apply to atomic gases like argon, neon and krypton.

First, let us focus on the concept of temperature. We can already make some
remarks based on these hypotheses. Within the framework of this theory, temperature
is the manifestation of the motion of molecules (H1). Moreover, the molecules move
with an average velocity calculated from the Maxwell–Boltzmann distribution (H2).
This average velocity is then used to calculate the kinetic energy (Greek: χινητιχός
"*kinetikos*", moving, putting in motion) for all the molecules. The faster the molecules
move, the greater the kinetic energy is. Therefore, since kinetic energy is related to
temperature, it follows that the velocity of molecules increases with temperature!

As for the concept of pressure, it is defined as follows. As we saw previously, the
best way to study a gas is to enclose it in a container. The walls of the container then
define a volume for the gas. As the molecules move ceaselessly and randomly (H1),
they hit the walls of the container very frequently, thus creating a pressure on the walls
(H3). The logical consequence is that, when the volume of the container decreases,
there are more and more collisions with the walls, and therefore pressure increases. On
the other hand, the larger the volume, the fewer collisions take place, and thus the pressure decreases. That's exactly what Boyle states in his law (PV = constant)! Similarly,
if temperature increases, the velocities of the molecules increase, as well as the number
of collisions with the walls. The momenta of the molecules ($\mathbf{p}=m\mathbf{v}$) are then larger, and
therefore the molecules hit the wall "harder" and more often. This means that, when
temperature increases, pressure increases. In this case, we recover Gay–Lussac's law
(P/T=constant)! If we imagine the walls to be "flexible", the increase in the number of
collisions with temperature will "push back" the walls, and thus volume will increase.
Simply put, if temperature increases, volume increases. It's Charles' law (V/T=con-
stant)! Finally, it is clear that, if the number of molecules increases, then the number
of collisions increases, which means that pressure increases! As we have seen before,
this implies that if the walls are "flexible" then volume increases. Here we find the so-
called Avogadro hypothesis (V/n=constant)! We therefore see that, if we use each of
these hypotheses, we find the famous law of ideal gas (PV=nRT)!

For those interested in the mathematical formalism, let us see how collisions
can be translated into equations. To simplify the calculation, we look at a single
molecule; the generalization to the entire system can be done by summing over all
molecules composing the gas. Moreover, for the sake of clarity, we only present
the results for an atomic gas (again, the calculation for molecules follows the same
reasoning, although it involves more complex mechanical equations). Prior to the
collision, the atom arrives towards the wall with the velocity $\mathbf{v}$. Then, after colliding
with the wall, it bounces back with a velocity $-\mathbf{v}$. As we can see, the norm $|\mathbf{v}|$ is con-
served, meaning that the velocity of the atom does not change in magnitude, but only
changes in direction during the collision. If we look at the momentum, before and

after the collision, we find that, initially, $\mathbf{p}_i = m\mathbf{v}$ and that, after the collision, $\mathbf{p}_f = -m\mathbf{v}$. The change in momentum is then $\Delta\mathbf{p} = \mathbf{p}_f - \mathbf{p}_i = -2m\mathbf{v}$. As we have seen previously, we know that a force can be calculated through the following equation, $\mathbf{F} = \Delta\mathbf{p}/\Delta t$ (Newton's law). It means that, in our case, if we look at the change in momentum during a time period of Δt, we can calculate the force exerted by the atom on the wall. Simply put, we have $\mathbf{F} = -2m\mathbf{v}/\Delta t$ which, in turn, gives an expression for the pressure on the wall $P = |\mathbf{F}|/A$, with A being the impact area on the wall. Finally, we find that the pressure due to the collision of one atom on the wall is equal to $P = 2m|\mathbf{v}|/(A\Delta t)$. Now, if we take into account the fact that, during the time period Δt, the distance traveled by the atom is 2L ("round trip") at the velocity $\mathbf{v}$, we have the relation $|\mathbf{v}| = 2L/\Delta t$. Noting that the volume $V = AL$, we find that $P = m|\mathbf{v}|^2/V$. In this example, so far, we have only taken into account a velocity, as well as a wall, along a single direction, i.e. $|\mathbf{v}_{1D}|^2 = v_x^2$. If we look at a 3D velocity vector colliding with 3D walls, there will be also components in the y and z directions, leading to $|\mathbf{v}_{3D}|^2 = v_x^2 + v_y^2 + v_z^2 \dots$, meaning that in our 1D example, $|\mathbf{v}_{1D}|^2 = 1/3|\mathbf{v}_{3D}|^2$. Plugging in these results, we obtain $P = m|\mathbf{v}_{3D}|^2/(3V)$. Then, if we generalize the results to N atoms, we end up with $P = Nm|\mathbf{v}_{3D}|^2/(3V)$. Finally, using the Maxwell–Boltzmann results for the root mean square velocity, we have $|\mathbf{v}_{3D}|^2 = 3kT/m$. Rearranging the equation, we get $PV = NkT$ or the famous ideal gas law $PV = nRT$ Q.E.D.!

This phenomenon of collision is an important element of modern physics. In fact, it is at the origin of the calculation of Avogadro's number by Einstein in 1905! It is indeed by using molecular kinetic theory, together with Brownian motion, that Einstein manages to calculate it. In 1909, Perrin definitively establishes the reality of atoms by estimating Avogadro's number through a series of experiments! It also leads to great mathematical advances pioneered by Wiener and Levy on the problem of "curves lacking tangents". They propose a model for which Brownian motion can be described with continuous paths, but with infinite instantaneous velocities! These discoveries are made possible by the famous botanist Brown (1773–1858) who, in 1827, identifies what is now known as Brownian motion by studying pollen through a microscope (see Figure 1.16). Indeed, he notices the presence of very small particles moving in all directions when pollen is immersed in water – the famous Brownian motion! He publishes his results in 1828 in "A Brief Account of Microscopical Observations Made in the Months of June, July, and August, 1827, on the Particles Contained in the Pollen of Plants; and on the general Existence of active Molecules in Organic and Inorganic Bodies".

As for the derivation of the kinetic theory of gases, many other hypotheses can be retained, leading to new results with more and more profound implications. For instance, if one takes into account the size of atoms/molecules, one can access transport properties, such as viscosity, diffusion and thermal conductivity. Then a question arises: What if we introduce interactions between atoms/molecules?

FROM THE IDEAL GAS LAW TO REAL GAS BEHAVIOR

The first piece of the puzzle is laid down by van der Waals (1837–1923), (see Figure 1.17). To account for the interactions between molecules (or atoms) in a gas, he starts from the ideal gas law ($PV = nRT$ or $PV = NkT$) and adds new terms. First,

FIGURE 1.16 Robert Brown among colleagues, including Cavendish, Dalton, Davy, Herschel and Rumford.

FIGURE 1.17 Johannes Diderik van der Waals (1837–1923).

he takes into account the size of the molecule and rewrites the definition for the volume. A molecule then occupies a volume b, which gives the new concept of accessible volume (V_a), defined as the total volume (V from the ideal gas law) minus the volume occupied by the N molecules composing the gas (N $\times$ b). This translates into $V_a = V - Nb$. Similarly, the presence of attractive interactions between molecules changes the definition for pressure. Indeed, pressure is defined as two terms. The first part is related to the pressure of an ideal gas, but with V_a instead of V. The second part arises from the attractive interactions and is modulated by a cohesive factor *a*. Altogether, we can write the pressure as $P = NkT/V_a - aN^2/V^2$. This gives us the famous van der Waals equation: $(P + aN^2/V^2)(V - Nb) = NkT$ or $(P + a'n^2/V^2)(V - nb') = nRT$. We note that *a* and *b* are molecule dependent, making this new equation able to describe real gases! For this extraordinary achievement, van der Waals receives the Nobel Prize in physics in 1910 for his equation of state, which successfully explains the gas and liquid states!

Let us add that these famous interactions are now known as van der Waals interactions. At first glance, they can be interpreted as an attraction that brings together two atoms or molecules, with an intensity that depends on the interatomic/intermolecular distance. More specifically, these interactions vanish when the distance becomes very large. In fact, it will be necessary to await the advent of quantum physics to fully understand this phenomenon.

One of the ideas that leads van der Waals to write his equation is the concept of compressibility (the ability for a gas or liquid to change its volume under pressure). Indeed, this property is poorly accounted for by the ideal gas model. The compressibility factor (Z) is therefore used to measure the deviation of a real gas from the ideal gas behavior. In other words, Z is the ratio of the real gas behavior ($P_{real}V_{real}$) with respect to the ideal gas behavior (PV from the ideal gas law, which can be replaced by nRT). Mathematically, $Z = (P_{real}V_{real})/(nRT)$. Plugging the compressibility factor into the van der Waals equation, we can rewrite the van der Waals equation into a polynomial equation of degree 3 in Z of the form $AZ^3 + BZ^2 + CZ + D = 0$. Historically, this is the first cubic equation of state. At a given temperature and pressure, solving this equation and finding a single real root makes it possible to identify if the fluid is either a gas or a liquid under these conditions. We can solve this equation graphically (when the equation $AZ^3 + BZ^2 + CZ + D = 0$ crosses the abscissa) or use algebraic methods pioneered by Cardano in the 16th century.

In his doctoral treatise in 1873, "*Over de continuiteit van den gas-en vloeistoftoestand*" (On the Continuity of the Liquid and Gaseous State), van der Waals also gives a description of the phenomenon of condensation and the concept of critical temperature. Indeed, he notices that each chemical substance has a critical point defined by a critical temperature, a critical pressure and a critical density. If the temperature and pressure of the experiment exceed the critical temperature and critical pressure of the substance, the substance is said to be supercritical! At this stage, the properties of the gas and liquid states are not different anymore, they are identical! We can no longer distinguish a liquid phase from a gas phase!

Van der Waals uses the critical properties to define new variables. He calls reduced temperature the ratio of temperature over the critical temperature, reduced pressure the ratio of pressure over the critical pressure and reduced volume the ratio of volume over the critical volume. And he observes that, for the exact same reduced properties (pressure, temperature and volume), all substances have the same compressibility factor! This gives the famous law of corresponding states! Thanks to this new law, we can obtain directly the gaseous and liquid properties of any substance. Since many scientists sought to liquefy gases at that time, van der Waals' work was of the utmost importance. Let us note that it is in 1908 in Leiden that Kamerlingh Onnes manages to liquefy helium below 4K (–270°C or –452°F), the coldest temperature achieved on Earth at that time!

When applied at the critical point, the law of corresponding states gives a reduced compressibility factor of $3/8 = 0.375$ for all substances. Nevertheless, some discrepancies have been found with the experimental results, showing that the law of corresponding states is not perfectly exact. Nowadays, we use another model known as a mean field approach.

Many different equations and models are now used to study gases and liquids. For example, the virial equation of state gives the compressibility factor as an expansion of $1/V_m$ (V_m = molar volume). Another example is the Peng–Robinson equation of state, developed in 1976 and now widely used in engineering to model natural gas systems in the petroleum industry.

2 Introduction to Thermodynamics
The Heat Concept

The acts of the mind, wherein it exerts its power over its simple ideas, are chiefly these three: (1) Combining several simple ideas into one compound one; and thus all complex ideas are made. (2) The second is bringing two ideas, whether simple or complex, together, and setting them by one another, so as to take a view of them at once, without uniting them into one; by which way it gets all its ideas of relations. (3) The third is separating them from all other ideas that accompany them in their real existence: this is called abstraction: and thus all its general ideas are made.

Locke, "An Essay Concerning Human Understanding", 1689

A QUESTION OF HEAT OR A MATTER OF TEMPERATURE

Is heat only a sensation? Heat is indeed a complex notion which has been difficult to define and to put into equations for a long time. This is nonetheless one of the sensations we apprehend best. Through our sense of touch, we can feel if an object is hot or cold. We are even capable of differentiating between something cold and frozen, as well as between warm and hot. The question is then how we can measure this sensation and put a number on its magnitude. Nowadays, we know from neuroscience that the sense of touch is not unique, but is a collection of distinct sensations, managed by two different systems. On the one hand, there is the epicritic system, which manages low-intensity stimuli, tactile sensibility and proprioception (perception or awareness of the position and movement of the body, sometimes called the sixth sense). On the other hand, there is the protopathic system, in charge of high-intensity stimuli and painful sensation. To differentiate between pain and temperature sensations, the human body utilizes different sensory receptors: nociceptors (Latin: *nocere* = to harm, a notion found in the Hippocratic Oath "*primum non nocere*": first, to do no harm) and thermoreceptors (Greek: *thermos* = hot). In particular, the latter can perceive changes in skin temperature of the order of 0.1°C! There are two types of thermoreceptors, those sensitive to warmth and those sensitive to cold, delimiting specific skin regions as warm or cold spots. Let us add that since humans are warm-blooded animals, we have many more cold receptors than warm receptors at the surface of the skin to help us adapt to the outer temperatures. Another cool fact

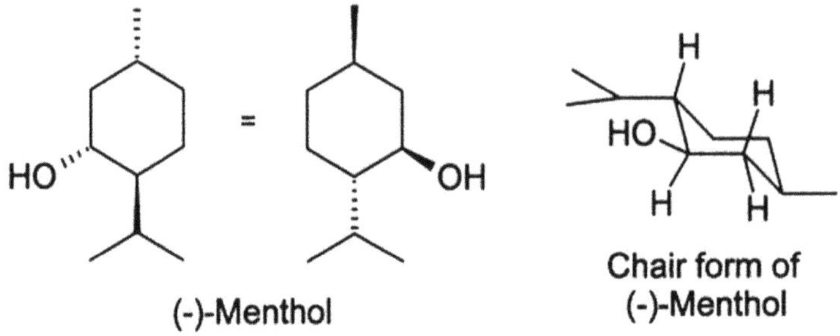

FIGURE 2.1 Chemical structure of menthol.

is that some thermoreceptors are polymodal, meaning that they can respond to warm and cold sensations and also to chemicals, such as menthol ($C_{10}H_{20}O$), contained in mint (see Figure 2.1), or capsaicin ($C_{18}H_{27}NO_3$), in cayenne peppers. Indeed, in the case of cayenne peppers, capsaicin activates the hot thermoreceptors, giving us the same sensation as we have with a hot cup of coffee! The same is true with menthol and the cold thermoreceptor, which creates the famous sensation of freshness that we find in gums and toothpastes.

Galileo is once again at the center of this question of heat with, around 1592, the invention of the thermoscope, from the Greek θερμός (*thermós* = warm, hot) and σκοπέω (*skopéō* = examine, inspect). It is based on an observation by Hero of Alexandria (1st century BC) that air expands as its heat increases. According to Castelli (1578–1643), Galileo's apparatus consists of a glass bulb attached to a long thin tube (see Figure 2.2). He begins by heating the glass bulb with his hands, turns it upside down and immerses the open end of the tube into a vessel of water. When he takes his hands off, water rises in the tube! He concludes that, when the air in the bulb gets colder, its volume decreases and water rises! By the same token, if the air gets warmer, its volume increases (it expands) and water falls down the tube! Galileo has just discovered an experimental device able to detect temperature changes! He then adds a scale to his device to allow him to determine if something is hot or cold. In 1623, Galileo publishes "*Il Saggiatore*" (The Assayer), in which he develops his thoughts on heat. In particular, he introduces the idea that motion is a cause of heat and that friction plays a significant role. Around 1650, Ferdinando II de Medici, Grand Duke of Tuscany, invents the first liquid-in-glass thermometer (similar to the ordinary thermometers we now use but without any reference points). To achieve this, he builds on Galileo's design and makes two small changes. First, the thin tube is now sealed, and second, the air in the bulb is replaced by a liquid, specifically here, alcohol (spirits of wine). This gives the first liquid thermometer also known as the "spirit thermometer". Let us add that, throughout the world, you can now buy a "Galileo thermometer". It is first designed at the end of the 17th century in Florence by a group of scientists mainly composed of pupils of Galileo. They are known as the "*Academia del Cimento*" (Academy of Experiments), and their motto is "*Provando e riprovando*" (Trying and trying again). Their "*termometro lento*" uses the fact

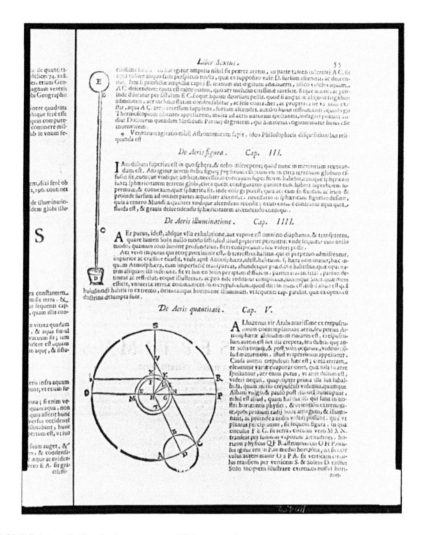

FIGURE 2.2 Galileo's thermoscope (17th century).

that the density of a liquid changes with temperature (one of Galileo's famous principles!). It is composed of a sealed glass tube filled with water. The thermometer also contains several floating glass bubbles of different densities, each carrying a metal tag indicating a specific temperature. The idea is that when the temperature of the air outside changes, the temperature of the water inside the tube also changes. Water then expands or contracts, modifying its density. As a result, the glass bubbles, depending on their density, can go up or down the tube as the temperature changes. Subsequently, one can read the temperature from the glass bubble that barely manages to float (or in other words, the lowest of those located at the very top of the thermometer).

At the same time, during the 17th century, scientists draw inspiration from Galileo's inventions and make tremendous progress. For example, in medicine, Santorio (1561–1636) develops, among other things, the pulse clock in 1602 and the first clinical thermoscope in 1612! Indeed, since Hippocrates and the early days of medicine, fever and chills have been deemed to be great indicators of bad health and death. However, only the hand is used to detect the patient's temperature, a subjective and sometimes misleading diagnosis. The development of the first mouth thermometer by Santorio is definitely a turning point towards modern medicine. It is an air thermometer, with a scale and numerical readings, that uses water as the indicator for temperature. Santorio's idea is to follow the evolution of his patients' health by monitoring their temperature. To do this, in the case of a fever, he carries out several numerical readings at different times. He then compares the results and concludes that, if the level of water in the thermoscope recedes, the patient's health improves. Differently put, the patient's temperature (and thus the fever) has decreased over time (the same explanation as with Galileo's thermoscope applies). On the contrary, if the level of water rises, the patient's health worsens. Let us add that each reading took about ten minutes to stabilize. However, Santorio's thermometer (as well as Galileo's thermoscope) is open to ambient air, and, as such, depends on atmospheric pressure, and thus on weather conditions!

Different temperature scales begin to appear during the 18th century. In 1701, Newton proposes his own scale (degree Newton: °N). It starts at 0°N (the temperature at which water starts to freeze), and we find among the principal points of the scale: 12°N, corresponding to the temperature of the human body, 24°N, corresponding to the heat of melting wax, 34°N when water "boyles vehemently", 48°N, corresponding to the melting point of equal parts of tin (Sn) and bismuth (Bi) and 96°N, corresponding to the melting of lead (Pb). In 1708, Fahrenheit (1686–1736) defines a new scale (degree Fahrenheit: °F). Incidentally, Fahrenheit's scale was inspired by a visit to Rømer (Danish astronomer). According to Fahrenheit, the main problem of Rømer's scale (°Rø) was its use of "inconvenient and awkward fractions". For this reason, he creates his own scale by first multiplying Rømer's scale by four and then by adjusting it again, ten years later. In 1714, he becomes the first to commercialize a mercury thermometer, allowing him to popularize his scale. Let us add that this is still the official temperature scale of the USA. Another important scale is the degree Réaumur (°Ré) introduced in 1730 by de Réaumur (1683–1757), which will be crucial for Laplace and Lavoisier's experiments on calorimetry. It is also still used nowadays in the cheese-making industry in Switzerland and Italy! Let us also note the degree Delisle (°D) proposed in 1732 by Delisle (1688–1768) in Saint Petersburg (Russia). It included up to 2,700 graduations to account for the harsh winters and remained in use for almost 100 years in Russia. Ten years later, in 1742, Celsius (1701–1744) invents a temperature scale (degree Celsius: °C) now commonly used in most of the world. His scale has two reference points: the boiling and freezing points of water. After his death, it is decided that the mark at 0°C corresponds to the freezing point and that the mark of 100°C is the boiling point of water. Note also that his scale is divided into 100 parts or marks (thus giving the name of centigrade to the Celsius scale). By the end of the 18th century, we thus have a measuring device, the thermometer, which allows for the quantification of temperature (or, in other words, assigning a number to different degrees of hot and cold). However, the question of what heat corresponds to remains an enigma.

The beginning of an answer comes with the identification of two different types of heat, with one of them not leading to an increase in temperature. Thanks to the discovery of the thermometer, Black makes extraordinary breakthroughs around 1760. Following a series of experiments, he defines a first type of heat known as specific heat. It represents the quantity of heat required to increase the temperature of a substance by 1°C per unit mass (unit: J/g/°C). In the case of water, one needs 4,180 J of energy to increase by 1°C an amount of 1,000 g (1 kg) of water, or, differently put, 1,000 calories of energy, approximately two slices of cheese pizza! He carries out another famous experiment where he studies the behavior of ice and water under ambient conditions. At the start of the experiment, Black puts some ice in a vessel (V_1) and some water in a vessel (V_2), both at a temperature of 0°C. After several hours, he notices that the temperature of V_2 has increased. On the other hand, the temperature of V_1 has remained at 0°C, while a little of the ice has melted! There is therefore a heat that is not detectable with a thermometer, which he calls latent heat (Latin: *latens* = lying hidden). It seems that it corresponds to an amount of energy (absorbed or released) by a substance during a constant temperature process (V_1 stays at 0°C throughout the experiment). Nowadays, it is written as the amount of energy required to completely achieve a phase change (here melting: ice → water) of a unit mass of a substance (unit: J/g). For example, in the case of water, one needs 334,000 J to transform 1 kg of ice into water at 0°C and ambient pressure. It means that, in Black's experiment, it would have required heating in a 700 W microwave oven for eight minutes to completely thaw the ice. By the way, the unit W, used for Watt and defining the power of a machine, was introduced by Black's pupil, James Watt.

FIRST THEORIES AND THE "ELEMENT OF HEAT"

The idea of the splitting of Nature into elements seems to come from Empedocles (490–430 BC) with the famous four classical elements: water, fire, air and earth. Associated with forces that he calls Love (a force that brings together) and Strife (a force that separates), those elements can form specific objects in Nature. For example, according to him, the main ingredients composing volcanoes are water and fire, as evidenced by lava. Let us add that, at that time, a theory based on "ἄτομον" (Greek: *atomon* = indivisible) is developed by Democritus (460–370 BC). Indeed, he states that everything cannot only be explained through what we see. There exists a very small world, impossible to see with the naked eye, composed of particles that cannot be split any further, which he calls "*atomon*". He thus introduces the idea that elements can be divided into atoma, giving the word "atoms" that we currently use.

A geometric explanation is proposed by Plato (428–348 BC) (see Figure 2.3) in "*Timaeus*". According to him, the four classical elements are made of regular solids! Fire can be represented by the tetrahedron (Greek: *tetra* = four; *hedron* = face), air by the octahedron (Greek: *octa* = eight; *hedron* = face), water by the icosahedron

FIGURE 2.3   Plato (428–348 BC).

(Greek: *icosa*=20; *hedron*=face) and earth by the hexahedron (Greek: *hexa*=six; *hedron*=face). The idea is that each face of these solids can be split into right-angled triangles, either isosceles or half-equilateral. These triangles can then be recombined to form any object! Let us note that aether will be proposed later as the fifth element and represented by the dodecahedron (Greek: *dodeca*=12; *hedron*=face), mimicking the shape of the Universe as a whole. Thanks to his theory, Plato is able to explain water's capacity to extinguish fire. Indeed, when water mixes with fire, we obtain air! Mathematically, if we count the triangular faces, water (20) + fire (4) = 3 air ($3 \times 8 = 24$)!

This concept of combination is then taken up by Aristotle (384–322 BC), one of Plato's pupils, in his famous cycle. For him, in addition to the four elements (Empedocles), there are also four elementary qualities: heat, cold, wetness and dryness. Indeed, each of the four elements is composed of a pair of elementary qualities. For example, we have for earth: cold and dryness, for water: cold and wetness, for air: heat and wetness and for fire: heat and dryness. Since each substance found in Nature is made up of the four elements, his idea is that the transformation of a substance into another can be achieved by changing the proportions between their elementary qualities. For instance, during combustion (heat), wood (earth) is transformed into flames (fire). Differently put, earth (cold and dryness) can be changed into fire (heat and dryness) if cold can be converted into heat through combustion. Aristotle's theory about transformations will influence greatly alchemy during the Middle Ages. Despite Democritus' atomic theory, people will choose to follow Aristotle's ideas. This may be because Aristotle was Alexander the Great's tutor

(Alexander III of Macedonia, 356–323 BC). Incidentally, Alexander the Great was one of the most successful military commanders and conquered, by the age of 30, a vast empire stretching from Eastern Europe to Asia and including parts of Africa.

An alchemist named Becher (1635–1682/1685) in *"Physica Subterranea"* (1669) reconsiders the theory of the four elements, developed during antiquity, and replaces them with three earth elements: *terra lapidea* (stony earth), *terra fluida* (liquid earth) and *terra pinguis* (oily earth). According to him, flames that appear during combustion are the manifestation of some *terra pinguis* being released into the air. He further proves his hypothesis by observing that ashes weigh less than the piece of wood prior to the combustion, demonstrating that something has escaped. His student Stahl (1660–1734) reformulates this theory in 1703 and gives a new name to *terra pinguis*: the phlogiston (Greek: φλογισ = inflame). According to him, all substances capable of burning (i.e. containing phlogiston) release phlogiston into the air during combustion, until the surrounding air becomes saturated (unable to absorb any more phlogiston). The formation of flames is interpreted as a proof that phlogiston is released. Moreover, the residue (or ashes), at the end of the combustion, is called a calx. In order to describe the combustion process, phlogistonists propose a first equation: substance → calx + phlogiston. Then, they turn to the phenomenon of calcination (strongly heating in air) for metals and propose a second equation: metal → (metal calx) + phlogiston. The concept of phlogiston remains valid until and during the 18th century, even if many experiments contradict it. One of the most famous objections comes from Lavoisier. Indeed, he shows that, instead of losing weight during combustion, some substances, including phosphorus (P), sulfur (S) and lead (Pb), gain weight!

Moreover, according to Lavoisier, there is conservation of heat! For this purpose, he defines heat as the "igneous fluid" or "matter of fire", which flows from hot bodies to cold bodies. In other words, the igneous fluid is composed of particles that are attracted to the particles of ordinary matter. For example, in water vapor, there is water (or ordinary matter) and igneous fluid. It allows him to characterize three states of matter, solid, liquid and vapor, depending on the amount of igneous fluid contained. For instance, a solid has little igneous fluid, whereas vapor contains a lot of it. In addition, building on Black's idea, he proposes the caloric theory, which, as its name indicates, deals with the phenomenon of heat transfer (Latin: *calor* = heat). According to him, absorption or release of heat can modify the properties of air. For instance, when air releases heat, it condenses. On the contrary, when air absorbs heat, it expands (as discussed in the previous chapter on ideal gases). Starting in 1783, Lavoisier carries out a series of experiments, together with Laplace (1749–1827), in which they use an ice calorimeter (see Figure 2.4) to measure amounts of heat (as discussed in their famous *"Mémoire sur la Chaleur"*). Indeed, Laplace designs a new experimental device, in which a sample chamber is surrounded by ice. Heat released from the sample melts the ice, which drips and is then collected to be weighted. We now have an apparatus capable of measuring heat: The first calorimeter is born! Moreover, they are able to show that there are two types of heat (*"chaleur"* in French), as suggested by Black; the first one is called *"chaleur libre"* ("free heat" or heat that flows from a hot body to a cold body) and the second one *"chaleur combinée"* ("combined heat" or heat that melts ice without inducing any temperature change). They observe that, for one pound of ice to melt, one needs a pound of

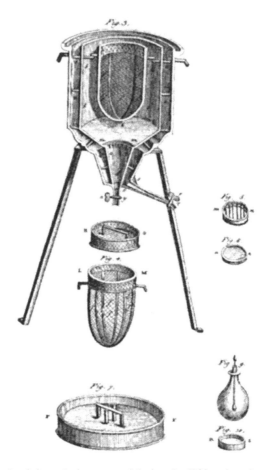

FIGURE 2.4 Laplace's ice calorimeter used during the 1783 series of experiments.

water at 64°Ré (degree Réaumur). Once both are mixed, there is no ice left within a matter of seconds, and all of the water is at 0°Ré. This shows that the amount of heat released when the temperature of a pound of water goes from 64°Ré to 0°Ré has been completely used to melt the pound of ice. The thermometer provides a proof of the existence of the "chaleur libre" or a heat flow from a hot temperature (64°Ré) to a cold temperature (0°Ré), giving one of the most used equations in calorimetry: $Q = m \times C \times \Delta T$. Here, 1 pound of water times C (heat capacity of water) times ΔT (0°Ré – 64°Ré) gives about $500 \text{ g} \times 4.186 \text{ J/g/°C} \times (0°C - 80°C) = -167{,}000$ J or –167 kJ. Note that, by convention, when heat is released, $Q < 0$ or $\Delta T < 0$ (this is why we have $\Delta T = 0°Ré - 64°Ré$). The process is named exothermic (Greek: *exo* = out of, *thermos* = heat). On the contrary when $Q > 0$ or $\Delta T > 0$, heat is absorbed and the process is called endothermic (Greek *endo* = inside, within). From our calculation, it means that one needs 167 kJ to melt a pound of ice. Now let us calculate the amount of energy needed to transform ice into water. For this, we use the formula for latent heat: $Q = m \times L_f$ where L_f is the latent heat of fusion. This gives about $500 \text{ g} \times 334 \text{ J/g} = 167{,}000$ J or 167 kJ, which is the exact opposite of the *"chaleur libre"*!

A proof of the conservation of heat! In other words, $mC\Delta T + mL_f = 0$! This means that the entire amount of *"chaleur libre"* generated by the change of temperature of water is completely transformed into *"chaleur combinée"* to melt the ice! Using the same device, Lavoisier and Laplace are the first to measure the heat released during chemical reactions, like the combustion of sugar, sulfur and phosphorus. They also calculate "animal heat" or the heat evolved by a guinea pig. Since respiration is a slow combustion, the consumption of *"air vital"* (oxygen) by the animal produces the *"air fixe"* (carbon dioxide) and creates heat. Incidentally, Laplace is Napoléon's professor at the *École Militaire* (Military School in Paris) and becomes his *"Ministre de l'Intérieur"* (Secretary of State). After six weeks, Napoléon replaces him because he "brought his idea of infinitesimals into public administration". It is only in 1824 that Clément (1779–1841) introduces the term "calorie" to denote the unit of heat. Nowadays, the calorie (cal) is the amount of energy or heat needed to increase 1 g of water from 14.5°C to 15.5°C under ambient pressure.

SUCCESSES AND FAILURES OF THE CALORIC THEORY

But the story does not end there. Thompson, Count von Rumford (1753–1814), calls into question the caloric theory in "An Experimental Enquiry Concerning the Source of the Heat Which Is Excited by Friction" (1798). By the way, he marries Lavoisier's widow in 1804 – Lavoisier being guillotined in 1794 in Paris during the French Revolution. In his book, Thompson explains that the real nature of heat is not caloric, but a form of motion: "Whence, then, came this heat? And what is heat actually? I must confess that it has always been impossible for me to explain the results of such experiments except by taking refuge in the very old doctrine which rests on the supposition that heat is nothing but a vibratory motion taking place among the particles of the body." These are the famous experiments on cannon boring. First, he observes that the tools used to bore a cannon are warming up and need to be cooled down by immersing them in water. The explanation at the time comes from the caloric theory. Indeed, there is a formation of metal chips during the boring of the barrel, and the caloric flows from these chips to the tools, making them hot. Following a close inspection of the chips, Thompson is not convinced by this explanation and decides to carry out another experiment. This time, workers use tools that are unable to form chips and place the boring device in water. The temperature of the water gradually increases and after two hours and 30 minutes, the water starts to boil without any formation of chips! This demonstrates that heat can be produced at will, and without limitation, simply by rubbing one object on another! This is a first observation that hints at a possible relation between heat and mechanical work! Another part of the answer comes from Laplace, for whom heat has a mechanical origin: "heat is the *vis viva* resulting from the imperceptible motions of the constituent particles of a body". To demonstrate this, he uses the calculation of the speed of sound. Since antiquity, it had been known that sound travels very quickly, but not instantaneously. For example, there is a time delay between the moment we see the flash of lightning and the moment we hear the sound of thunder. Starting in the 17th century, many scientists attempt to measure the speed of sound using, for example, the time delay between the flash and the sound of a firing cannon (Mersenne). In

1686, Newton uses the echo phenomenon to provide an estimate for the speed of sound (v). His idea is to use the echo of a clap of hands in a corridor to get an estimate by v=d/t. Knowing the distance (d = 128 meters) of the corridor, the only difficult task is then to calculate the time elapsed (t). To determine this time delay, he uses several pendulums with different lengths, until he finds one that has a swing period equal to the echo's return (5.3 cm!). He finds a value of 0.46 seconds. It is then easy to deduce a velocity: $c_{Newton} = 128/0.46 = 280$ m/s. Nevertheless, over the years, time measuring devices become more and more accurate and yield an experimental estimate of 330 m/s. The first to find the correct value is Laplace. He builds on Newton's intuition that the sound does not behave like a particle moving at very high velocity but propagates through successive vibrations of air particles! For him, sound travels through successive compressions and rarefactions of air, leading to fluctuations in temperature. Indeed, during a compression step, temperature increases while during a rarefaction step, temperature decreases. The steps take place so quickly that there cannot be any heat exchanged with the surroundings leading to what is called an adiabatic process! In 1816, he finally finds that "the speed of sound is equal to the product of the speed given by the Newtonian formula, by the square root of the ratio of the specific heat of the air under a constant pressure, and of its specific heat under a constant volume" giving rise to the famous formula for the speed of sound: $c_{Laplace} = \sqrt{\gamma} \; c_{Newton}$ with $\gamma = C_p/C_v$ (γ is known as the coefficient of Laplace). By replacing $\sqrt{\gamma}$ by its value $\sqrt{(7/5)} = 1.18$, he finds that the speed of sound is $280 \times 1.18 = 330$ m/s! Let us also add that this discovery wouldn't have been possible without Poisson, who develops the first mathematical equations for adiabatic processes. In particular, he develops the well-known pV^{γ} = const., to be compared to its isothermal (constant temperature) counterpart PV = const. (Boyle's equation). Let us add that nowadays supersonic aircraft can fly faster than the speed of sound, producing shockwaves and sonic booms (see Figure 2.5)!

But let us go back to the famous caloric. A very interesting experiment is performed by Clément (the same person who introduces the name calorie) and his father-in-law Désormes (1777–1862). Their idea is to capture the caloric in order to demonstrate its existence. They publish their results in the "*Journal de physique,*

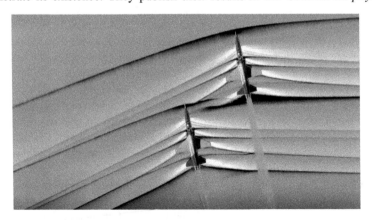

FIGURE 2.5 Shockwaves created by supersonic aircrafts.

de chimie, d'histoire naturelle et des arts" (Journal of Physics, Chemistry, Natural Science and Arts") in 1819. Here is their experiment. They use an air-filled glass vessel of a volume of 28.4 L at a temperature of 12.5°C and a pressure initially at 760 mmHg. Using a vacuum pump, they induce a depression of 14 mmHg, as indicated by a manometer. It reveals a pressure for the air inside the glass vessel of 746 mmHg. Then, they open a valve to let some air inside the glass vessel until the pressure inside is again 760 mmHg (atmospheric pressure). As soon as they reach atmospheric pressure, they quickly close the valve. To their surprise, they notice that the temperature inside the vessel has increased by 1.3°C, reaching a value of 13.8°C! This phenomenon is called adiabatic compression. Indeed, this is a compression (pressure increases from 746 mmHg to 760 mmHg) accompanied by a change of temperature (increase, in this case from 12.5°C to 13.8°C). Let us add that Clément and Désormes interpret the temperature rise as the capture of some caloric from the outside. They also find that "the absolute caloric from the outside, at the temperature of 12.5°C can raise by 114°C the same volume of air at atmospheric pressure and initially at the same temperature of 12.5°C". Let us note that this experiment shows that adiabatic processes are ubiquitous in Nature and not just related to the speed of sound. Indeed, we will see later that these processes play a key role in the development of engines, leading to the Industrial Revolution.

However, a question remains: How can one determine specific heats? Several measurements of this quantity for air are carried out at the end of the 18th century by Lavoisier, Laplace and Crawford. Some discrepancies are noted between the results. Crawford carries out his measurements in vessels with a fixed volume (C_v), while Laplace and Lavoisier determine the specific heat of air using a calorimeter at constant pressure (C_p). It is thought at the time that only the specific heat at constant pressure is of significance! Let us add that Delaroche and Bérard perform a series of experiments to determine the specific heat of many gases at constant pressure. They obtain a specific heat for air within 10% of the value we now know! Despite their initial assessment that the heat capacity at constant volume is of little significance, it turns out that both types of heat capacity actually play a major role in chemical processes and, most notably, in the determination of the speed of sound since $\gamma = C_p/C_v$.

THE "DUTCH GOLDEN AGE" AND THE IDEA OF AN ENGINE

In another part of Europe, in what we call nowadays the Netherlands, a new era begins. Under the rule of the King of Spain, and under duress because of his politics, high taxes and religious intolerance, the provinces revolt. To fight more efficiently against the Spanish army, the provinces sign the "Union of Utrecht" in 1579, and then the "*Acte de La Haye*" in 1581, which proclaims the independence of the "United Provinces". During the "Eighty Years' War" (Dutch: *Tachtigjarige Oorlog*) with Spain, the United Provinces assemble a Navy that rapidly becomes a threat to Spanish interests around the world. After the war, the Dutch Navy is converted into a merchant navy that takes over the international trade established by Spain, in particular in Asia and America. Speaking of which, in America, New Amsterdam (Dutch: Nieuw-Amsterdam) is the name of the Dutch settlement that grew up on Manhattan

Island. It is established by the Dutch West India Company and serves as the administrative capital of the New Netherland colony. As a result of the three Anglo–Dutch wars, New Amsterdam becomes New York in 1664 in honor of James, Duke of York (England), and New Netherland becomes part of New England. From the end of the 16th century until the beginning of the 18th century, the "Dutch Golden Age" (Dutch: "*de Gouden Eeuw*") enlightens Europe. Thanks to its commercial power as well as its emphasis on freedom, and most particularly, freedom of worship, this new federal republic attracts many people. Specifically, writers and scholars arrive as publishing houses are founded in Amsterdam and Rotterdam, together with the University of Leiden in 1575, creating a new center of knowledge in Europe. The "Golden Age" is often referred to in arts and culture since it is the cradle of a new movement in Dutch painting during the 17th century with masterpieces by Vermeer (1632–1675) (see Figure 2.6), Rembrandt (1606–1669) and Hals (1582/1583–1666).

One of the most famous Dutch families is the Huygens. Constantijn Huygens (1596–1687) is a diplomat and advisor to the house of Orange-Nassau. He is the secretary to two Princes of Orange, Frederick Henry, son of William I, and William II. Let us add that William I of Orange (William the Silent) organizes the Dutch revolt against Spain and that, later on, after 1815, the Netherlands becomes a monarchy under the House of Orange-Nassau. Huygens is one of the last Dutch Renaissance polymaths and has a great influence in different fields including diplomacy, music (lute), poetry and science.

FIGURE 2.6   "Young Woman with a Pitcher", ca. 1662 by Johannes Vermeer.

Twee poorten zeggen meer: onthaal ik vriend of gast
das is't niet door één deur, 't is door twee open deuren
De ruime ingang toont wat binnen zal gebeuren,
En dat de vrienden op mijn brood en op mijn wijn,
Niet half, niet heel, niet één-, maar twemaal welkom zijn.

Two Gates speak of more: be I greeting friend or guest
'tis not by one door yet by two open doors;
the spacious entrance shows what is to happen indoors,
And that friends are welcome to my bread and wine,
Not half, not whole, not once, but twice.

Constantijn Huygens, *"Hofwijck"*, 1653

Huygens maintains a voluminous correspondence and cultivates many friendships with renowned intellectuals of the time, including Descartes, Mersenne, Bacon, Rembrandt, Rubens, Corneille and Galileo; this has definitely a great impact on his son, Christian. Indeed, Christian Huygens (1629–1695) quickly becomes one of the great scientists of this era, as well as a major figure in the Scientific Revolution. Huygens makes groundbreaking discoveries in physics, most particularly in optics and mechanics. His most famous invention is the pendulum clock in 1658, a breakthrough that provides the most accurate timekeeper for centuries. As an astronomer, he is remembered for determining the true shape of the rings of Saturn and for the discovery of its satellite Titan (largest moon of Saturn). These discoveries are enabled by his design of an improved telescope and by the introduction of the Huygens' eyepiece. Let us add that Spinoza, the well-known philosopher, prepares his lenses. Finally, the NASA mission that lands on Titan on January 14, 2005 is named Cassini–Huygens to celebrate Huygens' discovery. Furthermore, a lunar mountain range (Mons Huygens) and a Mars crater, as well as the asteroid (2801) Huygens, are also named in his honor.

To promote the progress of Science in France, Colbert creates the "*Académie Royale des Sciences*" in 1666, in response to the "*Accademia dei Lincei*" in Roma (founded in 1603 by Federico Cesi and which helps Galileo during his trial) and the "Royal Society" in London (1660). Huygens is one of the first seven members of this Academy with Auzout (astronomer and founding member of the Paris Observatory), Roberval (mathematician and inventor of the Roberval balance, a weighing scale with two horizontal plates, that remained popular for three centuries), Carcavi (mathematician and keeper of the Royal Library), Frénicle (mathematician and counselor at the "*Cour des Monnaies*"), Picard (astronomer and in charge of calculating the Earth's radius by the Academy) and Buot (mathematician, astronomer). Huygens quickly assumes leadership of this new Academy. Among Huygens' assistants, there is a 23-year-old named Papin (1647–1712). His work consists of building vacuum pumps and designing new apparatus using combustion and explosion. In particular, Papin is very impressed by the so-called "Huygens' engine" and writes about it in 1675 in "*Nouvelles expériences du vide, avec la description des machines qui servent à le faire*" (New Experiments on the Vacuum, with the description of the machines used to do it). The idea is to use gunpowder to create a small explosion in a cylinder filled with air. As a result, the air expands, pushing up the lid on the top of the container, and then condenses as it cools down, bringing down the lid. If one attaches a weight to the lid with a rope, this means that such a device can lift this weight! This is the blueprint for a machine working with a combustion engine. However, because of Huygens' poor health, as well as the growing anti-Calvinist sentiment in France at the time, Papin moves to England.

In 1675, Papin becomes Boyle's new assistant thanks to a recommendation from Huygens. He publishes in 1681 the famous "A New Digestor or Engine for Softening Bones", which gives a description of a new device called a digester, now known as the pressure cooker! Indeed, this device uses steam to soften bones and, as such, is often referred to as a "digesting engine" (see Figure 2.7). The digester consists of a tightly closed container filled with water. The container is then heated up by fire and water turns into steam. This means that there is a very high pressure inside the container since it is tightly closed and thus the steam cannot expand beyond the container volume. To avoid any explosion, he puts a safety valve that only allows for a small fraction of the steam to escape when the maximum limit for pressure is reached. By doing so, the boiling point of water is raised considerably. Indeed, at ambient pressure, the boiling point for water is 100°, while it is 152° for a pressure of 5 bar. Thanks to this device, food can be cooked thoroughly much more quickly and using less energy (or wood at the time)! In 1705, after receiving a letter from Leibniz describing Savery's steam engine, Papin revisits Huygens' engine and starts thinking about replacing gunpowder with steam. He publishes his work in "*Ars Nova ad Aquam Ignis Adminiculo Efficacissime Elevandam*" (The New Art of Pumping Water by Using Steam) in 1707. He even builds a paddle-wheeled boat in 1709!

THE ADVENT OF STEAM MACHINES

One of the first steam-powered machines seems to be the "*aeolipile*" (Latin: *pila*=ball and Greek: Αἴολος Aiolos, keeper of the winds in Greek mythology) built

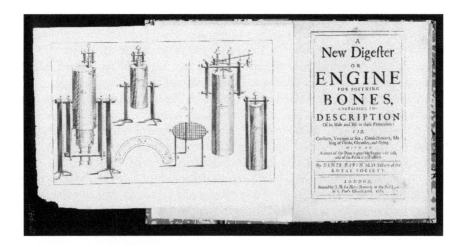

FIGURE 2.7 Papin's digester.

by Heron of Alexandria in the 1st century AD. It is the first steam turbine! The idea is to use the steam produced by a tank filled with hot water and heated by a fire. The steam is then collected by pipes, that are connected to a sphere. Two other bent pipes depart from this sphere to let the steam out and, by propulsion, make the sphere rotate (see Figure 2.8). It seems from the ancient manuscripts that this type of machine is used to automatically open and close the doors of temples! Until the 17th century, it remains mainly used for entertainment in "salons" in Europe. Let us recall that "salons" is the term used to describe gatherings, organized by women of high stature in society (baronesses, viscountesses, countesses, marchionesses, duchesses, princesses), in which participants engage in conversation about philosophy, sciences, literature, arts and politics. For instance, this is the place to be for scientists to present their new ideas and seek patronage. One of the most famous "salon" organizers is Catherine de Vivonne, Marquise de Rambouillet (Marchioness of Rambouillet). She hosts a mostly literary "salon" in her "Hotel of Rambouillet" (close to the Louvre and the Tuileries, residence of the King of France). Here, the art of conversation, wordplay and language refinement are central. For this reason, the participants are also called the *"Précieuses"* (precious). The French writer Molière

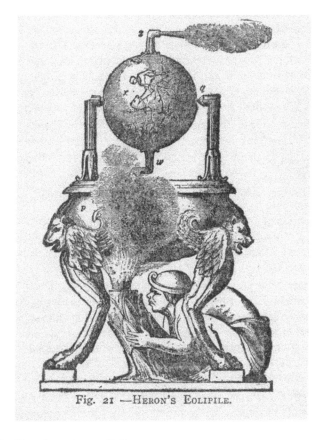

Fig. 21 —HERON'S EOLIPILE.

FIGURE 2.8 Heron's engine: aeolipile.

(1622–1673) writes a biting comedy of manners about them called the "*Précieuses ridicules*" (The Affected Young Ladies). The first representation is given in 1660 and is a great success. From then on, Louis XIV becomes Molière's patron. As for the operation of the aeolipile, it is necessary to wait for Segner (1704–1777) and the 18th century. Thanks to Newton's principle of action–reaction, Segner finally understands why the sphere rotates. Indeed, the ejection of steam towards the rear gives, by reaction, a thrust to the sphere, of equal force but opposite direction, hence going forward and making the sphere rotate! Differently put, due to its pressure, the steam ejected from the bent pipes causes the rotation of the rotor (sphere). It is exactly the same principle that is used nowadays in many devices like rotating sprinklers. Nevertheless, since wood is the only way to create heat at that time, and since heat loss is very significant, the Industrial Revolution is not possible yet.

In 1712, Newcomen (1664–1729) and Savery (1650–1715) build the first atmospheric steam engine, or "fire pump", and solve the issue of flooding in the Cornwall mines (England). Indeed, the evacuation of infiltration water is a major issue, and the flooding of the wells prevents the extraction of ore beyond a specific depth. Newcomen and Savery's intuition consists of using thermal energy from the heat of a fire and transforming it into mechanical energy to operate water pumps. The device can be described as follows (see Figure 2.9). It consists of a large horizontal beam connected on one side (right) to a cylinder (steam piston), and on the other side (left) to a water pump. The idea is to use a boiler to create steam and a water tank to provide cold water. In practice, the beam rocks to the right when cold water is injected, and rocks to the left when steam is let in. The repeated rocking is the heart of the machine that pumps water from the mine. From an engineering standpoint, the counterweight on the left is down, pulling up the piston in the cylinder. The steam produced by the boiler goes into the cylinder via a pipe. Then the pipe is closed thanks to a valve. Cold water from the water tank is injected into the cylinder through another pipe. This causes the steam to condense in the cylinder pulling down the piston and thus rocking the beam to the right. Indeed, the pressure within the cylinder is now less than the atmospheric pressure. Thus, the atmospheric pressure pushes down the piston, giving this machine the name "atmospheric steam engine"! As the beam rocks to the right, water from the mine is extracted through the water pump. Then, the valve on the pipe is opened again, letting steam in the cylinder and the counterweight goes down again, completing the cycle for the engine. And so on. Let us add that in its original version, the shut-off valve was operated manually. Newcomen engines are extremely robust, reliable and can operate day and night. This makes them extremely successful and prefigures the Industrial Revolution. When Newcomen passes away in 1729, there are more than 100 of his engines working throughout the UK and Europe. These engines remain in use throughout the 18th century.

The advent of the machine era is deeply connected to the empiricist philosophy introduced by Bacon in England at the turn of the 17th century. Indeed, the development of new machines to address practical issues is firmly rooted in the experiment-driven discovery process advocated by Bacon. Locke (1632–1704) is an empiricist from the modern era. According to him "No man's knowledge here can go beyond his experience". In his most famous book, "An Essay Concerning Human Understanding", he goes further by addressing the thought-provoking question:

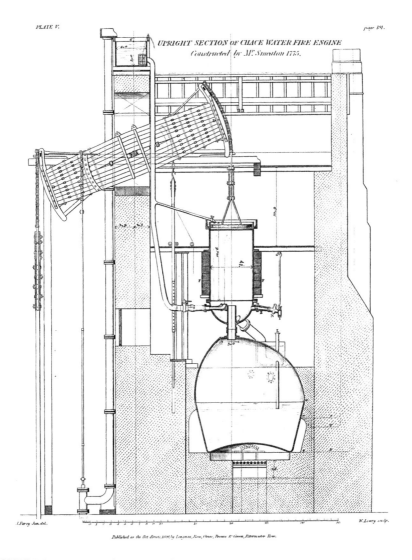

FIGURE 2.9   Newcomen's steam engine.

What is the capacity of the human mind for understanding and knowledge? Indeed, he posits that all objects of understanding are ideas and that these ideas exist in the mind. One of his objectives in this book is thus to trace the origin of ideas. According to him, innate ideas, imprinted from birth in the mind, do not exist: All human knowledge comes from sense experience first, and then from reflection or thinking. He also categorizes ideas into "simple ideas" and "complex ideas". Simple ideas are those that come directly from the senses. They are "simple" and can serve as raw material for further thinking. Complex ideas originate from combinations of simple ideas, made by the mind using abstraction. "When the understanding is once stored with these simple ideas, it has the power to repeat, compare, and unite them,

even to an almost infinite variety, and so can make at pleasure new complex ideas."
Locke also studies the concept of sameness through the sameness of a person, the
sameness of soul and the sameness of body, giving rise to the first theory of personal
identity. Indeed, he suggests that everything hinges on the continuity of consciousness. In other words, he thinks that the sameness of a person with time is not only
about preserving the same body and the same soul, but also in having the same
series of psychological connections: memories! For Locke, "to be a person is to be
an intelligent thinking being who can know himself as self, the same thinking thing
in different times and places", maybe a concept that will be key in our future with
the advent of machine learning and artificial intelligence.

As for Berkeley (1685–1753), the important question is: What allows an idea to represent a material object? To answer this question, he proposes the likeness principle in
his book "A Treatise Concerning the Principles of Human Knowledge" (1710): "But say
you, though the ideas themselves do not exist without the mind, yet there may be things
like them whereof they are copies or resemblances, which things exist without the mind,
in an unthinking substance. I answer, an idea can be like nothing but an idea; a colour
or figure can be like nothing but another colour or figure." According to him, one cannot think of mind-independent objects since one must employ a thought to represent an
object. This leads to a new philosophical theory, known as immaterialism, later referred
to as idealism: Objects, such as books and tables, are only ideas in the minds of perceiv-
ers and thus cannot exist without being perceived! Let us add that Berkeley's motto was
"esse est percipi (aut percipere)" –(to be is to be perceived (or to perceive)).

David Hume (1711–1776) is one of the most important philosophers of the 18th century.
His books "A Treatise of Human Nature", "Enquiries concerning Human Understanding"
and "Concerning Principles of Morals" continue to draw attention, impacting Kant,
Comte, Mill and many philosophers from the 20th century. His work draws from the
scientific advances made by Newton, as well as from the empiricism proposed by Locke.
Hume throws doubt on the principle of causality central to Newton's scientific advances:
"Our idea, therefore, of necessity and causation arises entirely from the uniformity observable in the operations of nature, where similar objects are constantly conjoined together,
and the mind is determined by custom to infer the one from the appearance of the other".
He believes that experience does not provide the supposed necessary connection between
cause and effect. In particular, the belief that an event A causes another event B can also
be interpreted as resulting from the closeness between the two events A and B in space
and time, without one being the consequence of the other. In other words, he proposes that
events can happen as coincidences without any causation. "We know nothing farther of
causation of any kind than merely the constant conjunction of objects, and the consequent
inference of the mind from one to another."

THE CONCEPT OF EFFICIENCY AND THE
INDUSTRIAL REVOLUTION

Another important inventor of this era is undoubtedly Watt (1736–1819) (see
Figure 2.10). He realizes that Newcomen and Savery's steam engine wastes most of
the energy provided by steam. Indeed, during the cooling process, the steam condenses, but the cylinder also cools down, meaning that energy will be needed during

FIGURE 2.10   James Watt (1736–1819).

each cycle to reheat the cylinder! His idea is to add a separate piston chamber or condenser, intended for the condensation of steam. It works as follows. As previously, a boiler produces steam that enters the cylinder. To make the beam rock to the right, the valve between the cylinder and the condenser is open, allowing steam to enter the condenser. Inside the condenser, cold water is injected and steam changes into water. By doing so, the pressure inside the cylinder becomes lower than the atmospheric pressure on the outside. The piston of the cylinder is thus pushed down, leading to the rocking of the beam to the right. Then, the valve is closed, and the cycle starts again: The boiler provides steam to the cylinder and so on. The tremendous advantage of this improvement is that there is no cooling of the cylinder anymore. It is immediately ready for another stroke! Watt continues to work on his machine and proposes a more reliable design that uses much less coal than Newcomen's engine to yield the same amount of power. In 1781, he introduces a system, using sun-and-planet gear, to transform the linear motion of the engines into rotary motion in order to replace the wheels used in mills! In 1782, Watt also invents the double-acting engine, in which the piston pushes and pulls, meaning that there is no more need of the counterweight! He develops the concept of horsepower (hp) to compare the output of steam engines with the power of draft horses. Let us note that, in honor of his work, the power unit is now written in Watt (W) with 1 hp = 746 W. It represents the unit of measure for the energy flow, i.e. 1 Joule of energy, transferred during 1 second. For example, the power of an electric engine (electric car) is of the order of 100 kW (1 kiloWatt = 1000 W=10^3 W), a nuclear reactor produces a power of the order of 100 MW (1 MegaWatt = 1,000,000 W = 10^6W) and a space shuttle requires a power of the order of 10 GW to take off (1 GigaWatt = 1,000,000,000 W = 10^9W)! At the end of the 18th century, the Industrial Revolution can begin!

One of the most important changes occurring in Great Britain at the time is the transformation of royal monopolies into patents. Indeed, a new law is voted so that any invention, that is both original and useful, can be patented. The great advantage of a patent is that, whenever someone uses the invention for commercial gain, the inventors receive money! Nevertheless, a patent is only valid for a limited time period, up to 14 years; after that, everyone can use the invention for free! As for the rules, a patent can only be granted if it represents a significant advance and not a minor change to a previously patented invention. This new law leads to a tremendous increase in trade and business that greatly contributes to the Industrial Revolution. A famous example is the story behind Watt's patent "A New Invented Method of Lessening the Consumption of Steam and Fuel in Fire Engines", granted in 1769. Let us add that his patent proposal is financed by loans from Black (the discoverer of latent heat). In 1772, Boulton, owner of the Soho Works Factory (Birmingham, UK), buys a share of Watt's patent and begins to develop its industrial applications. Together, they produce a multitude of new steam engine designs. The expansion is extremely rapid!

The Industrial Revolution also marks the transition from manual labor to the advent of machines and the factory system. By the mid-18th century, the UK is the world's commercial leader, with a global trading empire including the famous East India company that trades goods with what is now India, Japan and China. The development of machines also requires a larger consumption of raw materials, such as coal (black sedimentary rock – coal originates from the decay of vegetable matter into peat which, over millions of years, converts into coal). The chemical equation for the combustion of coal is $C + O_2 = CO_2$, releasing heat in the process (for example, used to form steam in Newcomen's and Watt's engines). In this equation, coal is represented by its major component carbon (C), which reacts with air (specifically oxygen: O_2) to form carbon dioxide (CO_2). The heat of combustion for coal is 33 kJ/g. For example, using a boiler from a steam engine, one will need to provide 2,400 kJ in order to transform 1 kg of water into steam (heat of vaporization of water: 2,400 kJ/kg). The amount of coal is then given by the ratio $2,400/33 = 73$, meaning that one needs 73 g of coal to produce 2,400 kJ of energy, which, in turn, is enough to transform 1 kg of water into steam. This shows how much the Industrial Revolution depends on the efficient transport of coal. At the time, this is achieved by boat. "If Britain had had to depend on her roads to carry her heavy goods traffic the effective impact of the Industrial Revolution may well have been delayed until the railway age" (Deane). So, it is no surprise that canals start to appear. One of the most famous examples is the Bridgewater Canal, built by Francis Egerton, 3rd Duke of Bridgewater, after having seen the "Canal du Midi" in France. In 1764, the Bridgewater canal reaches Manchester (UK), and finally links Manchester to the Mersey estuary (near Liverpool, UK) in 1776, giving access to the Irish Sea. This canal leads to a highly efficient and cost-effective transport of coal, which allows for the rapid growth of cities in the north of England. To give an idea, horse-drawn barges were still used on London's canals until the 1960s. On the sea, steamboats evolve rapidly. In 1840, the first steamer carrying passengers, or ocean liner, the RMS (Royal Mail Steamer) "Britannia" (see Figure 2.11) sails from Liverpool (UK) to Boston (USA) in 12 days and 10 hours! Incidentally, Dickens makes his first visit to America in 1842 on the RMS "Britannia". Starting in 1910, the "Blue Riband" is awarded to the ocean liner that crosses the Atlantic Ocean the fastest. The RMS "Mauretania" sets

FIGURE 2.11 RMS Britannia.

the record to roughly five days in 1909! This record remains unbroken until 1929. On the land, there is a tremendous railway boom. The first railway links Stockton with Darlington over an 8-mile track in 1825. This allows the efficient transport of coal and decreases its price by 33%! In 1830, the "Liverpool and Manchester railway" is the first steam-powered railway to transport both passengers and freight between two cities and to allow passengers to make return journeys within the same day! In 1830, there are 125 miles of railway lines in Britain. This number is multiplied by roughly 100 to reach more than 13,000 miles in 1871! To operate these new machines, charcoal is used rather than wood, since it contains more carbon and thus, produces much more heat for the same weight. Later, it is replaced by coke, which has an even higher carbon content and thus yields more energy. This will play an important role in the fabrication of steel since it is obtained by mixing coke and iron in a blast furnace (1,600°C). Indeed, iron is known to be a soft metal and is not strong enough to be used in machine designs, while steel is. Numbers from 2013 show that, to this day, steel remains the most important engineering metallic material with 1.6 billion tons produced in the world, 34 times more than the second most important, aluminum!

Finally, the Industrial Revolution is also characterized by a demographic boom. In 1750, the world's population is 700 million and grows by more than 40% in 50 years to reach 1 billion in 1800. More and more inventions change people's lives. In 1836, Morse devises the telegraph. Forty years later, in 1876, Bell invents the telephone. Edison designs the phonograph in 1877 as well as the first record in the world! Finally, the Wright brothers build the first airplane in 1903 and realize the myth of Icarus and Daedalus: to fly!

BIRTH OF A DYNAMIC THEORY OF HEAT OR THERMODYNAMICS

Carnot (1796–1832) becomes famous for his memoir *"Réflexions sur la puissance motrice du feu et sur les machines propres a déveler cette puissance"* (Reflections on the Motive Power of Fire) published in 1824. With his engineering background, he is fascinated by the operation of steam engines. Most recent advances rely on improving the materials used to build these engines. For Carnot, it is possible to develop a theory based on mathematical calculations that can guide further improvements and lead to more efficient machines. In his book, he proposes a machine theory that can explain the relationship between the flow of caloric (heat transfer) and the motion of the parts of the engine. "Everyone knows that heat can produce motion. That it possesses vast motive power no one can doubt, in these days when the steam engine is everywhere so well known". This is different from Newton's mechanics, in which a force needs to be exerted on an object to create motion. For this, Carnot builds on Lavoisier's caloric theory and analyzes the flow of caloric from a hot body to a cold body, which, according to him, is the essence of the operation of steam engines. Indeed, he observes that the motive power rises from the drop of heat from the high temperature of steam to the low temperature of the condenser, meaning that work can be produced by heat! "The production of motion in steam engines is always accompanied by a circumstance on which we should fix our attention. This circumstance is the re-establishment of equilibrium in the caloric; that is, its passage from a body in which the temperature is more or less elevated, to another in which it is lower." More specifically, he states that there is no consumption of the caloric during combustion, but only transport of the caloric from a hot body to a cold body. "The caloric developed in the furnace by the effect of the combustion traverse the walls of the boiler, produces steam, and in some way incorporates itself with it. The latter carrying it away, takes it first into the cylinder, where it performs some function, and from thence into the condenser, where it is liquefied by contact with the cold water which it encounters there. Then, as a final result, the cold water of the condenser takes possession of the caloric developed by the combustion. It is heated by the intervention of the steam as if it had been placed directly over the furnace. The steam is here only as a means of transporting the caloric." Or, in other words, the production of engine power is due to the presence of a hot body, but also of a cold body! It means that it only takes a difference in temperature to produce motive power! Based on this principle, he proposes an exercise of thought nowadays known as the "Carnot cycle". It makes it possible to build an ideal steam engine, i.e. by using a cycle (or successive operations) as efficient as possible to produce the maximum amount of work from heat. Since this is a thought experiment, various elements of the engine (insulating walls, no friction of the piston) operate without any loss of heat, which, of course, is not possible in real life. He summarizes these results as follows: "The motive power of a waterfall depends on its height and on the quantity of the liquid; the motive power of heat depends also on the quantity of caloric used, and on what may be termed, on what in fact we will call, the height of its fall, that is to say, the difference of temperature of the bodies between which the exchange of caloric is made". The Carnot cycle thus allows the calculation of the engine efficiency (%) by the following formula: $(T_H - T_C)/T_H \times 100$, where T_H is the temperature of the hot body and T_C the temperature of the cold body. The wider the temperature range, the more efficient the cycle!

This observation is very important and draws significant interest most particularly in the engineering community. Among them, Clapeyron (1799–1864), an engineer from a French railroad company (see Figure 2.12), is put in charge of the construction of the railway line between Paris and Saint-Germain-en-Laye. Incidentally, Louis

FIGURE 2.12 "Arrival of the Normandy Train, Gare Saint-Lazare", by Claude Monet (1877).

XIV (King of France who built the Chateau de Versailles) was born in 1638 in Saint-Germain-en-Laye. Clapeyron is naturally interested in Carnot's work on engine power and decides to follow up on Carnot's ideas. He develops a mathematical representation or graph to plot the evolution of pressure (P) as a function of volume (V). This PV diagram is so fundamental to the birth of thermodynamics that it is nowadays also called "Clapeyron's diagram". It allows the translation of experiments into mathematical objects. For example, a simple experiment (single process) can be described by a line and a cycle (several successive processes) into a parallelogram. This tool allows him to make a new interpretation of the Carnot cycle in "*Mémoire sur la puissance motrice de la chaleur*", published in 1834, and to generalize the process to any gas (see Figure 2.13). It consists of four steps. First, the gas inside the cylinder is in contact with the steam from the boiler at temperature T_1. The gas starts to expand. This first step is then called a reversible isothermal expansion. Second, the system is no longer in contact with the boiler and the gas continues to expand without any heat exchange with the surroundings. This is a reversible expansion process, and since there is no heat exchanged (ideal process), we call it a reversible adiabatic expansion. Third, the gas is in contact with the water from the condenser at T_2 with $T_2 < T_1$. The gas is then cooled down and starts to undergo compression at temperature T_2. This process is therefore a reversible isothermal compression. Fourth, the system is no longer in contact with the condenser and continues to undergo a compression without any heat exchange. This process is a reversible adiabatic compression. These processes constitute the Carnot cycle. If we follow the direction of the successive processes, we can see that we turn in a clockwise (anti-trigonometric) sense. It means that the cycle produces work! This is

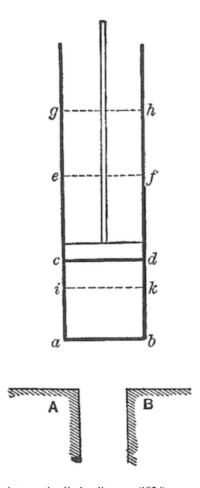

FIGURE 2.13 Carnot's piston and cylinder diagram (1824).

exactly what Newcomen and Watt had found in their experiments! Moreover, Clapeyron goes further and uses analytical mathematical formulae to describe this cycle. First, for the reversible isothermal process, the gas is modeled by an ideal gas, meaning that the variation of pressure (P) and volume (V) can be known through the famous ideal gas law: PV=nRT. In the PV plane (P as a function of V or mathematically P(V)), the equation P(V) = nRT/V is a hyperbola (a curved line between pressure, on the vertical axis, and volume, on the horizontal axis). We thus have a curved line that goes towards the right of the PV plane. Let us add that, as temperature increases, the plot for this line is farther away from the origin. Second, the reversible adiabatic expansion. Since the gas is assumed to be an ideal gas, we can use the Poisson equation for an ideal gas PV^γ = const. Let us add that, at this time, Clapeyron is not aware of this formula, as it does not appear in his paper. In the PV plane, P=const./V^γ corresponds to a power law curve. Since for any gas $\gamma > 1$ (for example, for air, $\gamma = 1.4$), the pressure in the PV plane decreases faster with respect to volume for an adiabatic process than for an isothermal process. Since we deal with an expansion, the volume increases, meaning that the power law curve decreases and goes towards the bottom right corner of the PV plane. Third, the reversible isothermal process. We have the same type of curve to describe this process as before. Here the curve heads towards the bottom left corner on the PV plane. Fourth, the adiabatic reversible compression. This follows the same type of curve as for the adiabatic reversible expansion. However, this time, this is a compression, meaning that pressure increases. The curve thus heads towards to the top left corner of the PV plane. In order to calculate the total work done by the system, one needs to calculate the amount of work done for each step. Indeed, each amount of work can be determined as the area defined between a curve and the horizontal axis. In other words, it corresponds to an integral function. For the first step, there is an increase in the volume of dV between V_1 (initial volume) and V_2 (final volume), we have

$$W_1 = \int_{V1}^{V2} nRT_1/VdV = nRT_1 \ln\left(V_2/V_1\right).$$

For the second step between V_2 and V_3 we have

$$W_2 = \int_{V2}^{V3} const./V^\gamma dV = const./\left(1-\gamma\right)\left(V_3^{(1-\gamma)} - V_2^{(1-\gamma)}\right).$$

For the third step between V_3 and V_4 we have

$$W_3 = \int_{V3}^{V4} nRT_2/VdV = nRT_2 \ln\left(V_4/V_3\right).$$

Finally, for the fourth step between V_4 and V_1, we have

$$W_4 = \int_{V4}^{V1} const./V^\gamma dV = const./\left(1-\gamma\right)\left(V_4^{(1-\gamma)} - V_1^{(1-\gamma)}\right).$$

The total sum gives

$$W = W_1 + W_2 + W_3 + W_4 = W_1 + W_3 = nRT_1 \ln\left(V_2/V_1\right) + nRT_2 \ln\left(V_4/V_3\right) > 0.$$

(To find that $W_2 + W_4 = 0$, just recall that the two adiabatic processes connect T_1 and T_2 but in opposite sense, leading to $V_3/V_4 = V_2/V_1$).

We can also understand the total work as the following (here, system = air). For W_1, since $V_2 > V_1$, $W_1 > 0$, meaning that the system does work as it expands. For W_2, since $V_3 > V_2$, $W_2 > 0$, meaning the system does work again since it continues to expand, but this time adiabatically. For W_3, since $V_4 < V_3$, $W_3 < 0$, meaning that, this time, the surroundings do work on the system. A gas does not undergo a compression on its own, it needs some help! For W_4, since $V_1 < V_4$, $W_4 < 0$, meaning that the surroundings still do work on the system to continue compressing the gas. Let us add that there is another convention (Europe), which stipulates that any energy that flows out of the system is counted as negative, while energy that flows into the system is counted as positive. In this convention, $W_1 < 0$ (the system does work, resulting in an energy flow out of the system), $W_2 < 0$ (same reason), $W_3 > 0$ (the system receives some help from the surroundings, resulting in an energy flow in the system) and $W_4 > 0$ (same reason).

Finally, if the cycle is that of an engine ($W > 0$), we can calculate the efficiency $\eta = (T_1 - T_2)/T_1$. Let us recall that the Carnot cycle represents the most efficient way to produce work from two sources of heat! The same calculation, using the European convention, leads to a cycle working as an engine if $W < 0$. Indeed, in this convention, $W = -\int PdV$ and takes into account the fact that energy flows out of the system.

However, as we will see later, the efficiency can also be calculated as the ratio of the work done over the heat supplied. Using the first law of thermodynamics: $\Delta U = Q - W$, we can calculate the heat required by the system to work. For an isothermal process, it gives $Q = W$, while for an adiabatic process $Q = 0$. For the first step of the cycle, the heat required is given by $Q_1 = W_1 = nRT_1 \ln(V_2/V_1)$. We can now calculate the efficiency.

$$\eta = \frac{\text{work done}}{\text{heat supplied}} = \frac{nRT_1 \ln(V_2/V_1) + nRT_2 \ln(V_4/V_3)}{nRT_1 \ln(V_2/V_1)}$$

$$= T_1 - T_2/T_1 !$$

Again, we reach the same conclusion using the European convention with $\Delta U = Q + W$.

Let us add that if we look at the total sum of work and heat, we end up with the contour integral $\oint_C W_{tot} + Q_{tot} = 0$, meaning that the system goes over a cycle, and thus comes back to exactly the point it started from. It also uncovers something that is crucial for thermodynamics, the concept of state functions. Such functions only depend on the point the system is at and not the path it takes to get there.

However, because of all the assumptions of ideality, it is not possible to achieve this efficiency in real-life applications: "To render useful all the motive force at our disposal, the detent should be continued until the temperature of the vapour is reduced to that of the condenser, but practical considerations, suggested by the manner in which the motive force of fire is employed in the arts, prevent the attainment of this limit".

Let us add that Clapeyron is also known for an equation bearing his name, later called the Clausius–Clapeyron relation. This time, Clapeyron concentrates his efforts on the PT plane. He succeeds in determining "phase diagrams", which delimit regions in the PT plane where the different states of matter (liquid, solid, vapor) exist. Thanks to this representation, it is thus possible to know if, for example, ice is converted into water for a given temperature and pressure. We can then easily define the changes in states, such as vaporization (transition from a liquid state to a vapor state), condensation (vapor to liquid), solidification (liquid to solid), fusion (solid to liquid) and sublimation (solid to gas). The Clausius–Clapeyron relation gives the slope of the tangents of this curve: $dP/dT = L/(TdV)$, where dP/dT is the slope of the tangent to the coexistence curve at any point, L the latent heat and dV is the specific volume change of the phase transition. Nowadays, latent heat is called enthalpy of change and is defined as the necessary quantity of energy for 1 mole or 1 kg of a pure body to change its state.

CONSTITUTION AND LAWS OF THERMODYNAMICS

Since Newton, mechanics is the favored approach to explain phenomena occurring in Nature. Basically, if something happens, it means that there is a cause – or, in other words, if there are no external forces acting on a system, nothing should happen. However, from Carnot's observations, it seems that heat passes from hot to cold without anything, or anyone, from the outside intervening to trigger this transfer. What happens is a mystery!

An answer comes from Joule (1818–1889). In the process of making a link between the effects of mechanical action and the resulting heat, Joule carries out an interesting experiment. The apparatus consists of two weights and of a paddle-wheel immersed in a barrel of water. Because of gravity, the weights become falling weights that spin the paddle wheel (see Figure 2.14). He quickly remarks that the water temperature inside the barrel increases. Indeed, he discovers that he needs to repeat the process 20 times to increase the temperature by about 1°F! In 1850, he publishes in "On the Mechanical Equivalent of Heat" a report on this experiment. He clearly demonstrates that gravity, acting as mechanical work, is transformed into heat and causes the water temperature to increase! "From this fact I considered myself justified in announcing that the quantity of heat capable of increasing the temperature of a pound of water by one degree of Fahrenheit's scale, is equal to, and may be converted into, a mechanical force capable of raising 838 pounds to the perpendicular height of one foot." Nowadays, 1 Joule (J) corresponds to the application of a force of 1 Newton (N) over a length of 1 meter (m). It can also be expressed in terms of power (W). Indeed, the power of 1 W is the amount of energy (1 Joule) expanded during one second.

FIGURE 2.14   Joule's heat apparatus (1845).

In 1847, Thomson (1824–1907) hears about Joule's findings during the British Association for the Advancement of Science meeting in Oxford. He is very interested, however skeptical, in Joule's ideas. Nevertheless, they start a very fruitful correspondence during the next few years, which leads to a great collaboration. Among others, let us mention the now famous discovery known as the "Joule–Thomson effect". Indeed, during an experiment, they find that when they flow high-pressure air through a small porous plug, the pressure drops, and the air is cooled down! Building on this work, in 1895, Linde and Hampson independently develop and patent a process that leads to refrigerators based on the Joule–Thomson effect. Let us add that this effect is at the center of what we now call cryogeny technology! Finally, Thomson is ennobled in 1892, becoming Baron Kelvin or Lord Kelvin – which is why we can find his discoveries under the name of either Thomson or Kelvin. This is the very same person! During that time, Clausius (1822–1888) introduces the notion of "mechanical theory of heat" or "thermodynamics" that aims at finding a common ground between Carnot's principle and the then well-established concept of conservation of energy. The theory, as well as the corresponding laws, are stated in his famous paper "*Ueber die bewegende Kraft des Wärme und die Gesetze, welche sich daraus für die Wärmelehre selbst ableiten lassen*" (On the Moving Force of Heat and the Laws regarding the Nature of Heat itself which are deducible therefrom) in 1850. The first explicit statement of what we now call the "first law of thermodynamics" is "In all cases where work is produced by heat, a quantity of heat is consumed proportional to the work done; and inversely, by the expenditure of the same amount of work the same quantity of heat may be produced". Using a mathematical formalism (following Clapeyron's work), he also introduces the concept of internal energy: "Work may transform itself into heat and heat conversely into work, the quantity of the one bearing always a fixed proportion to that of the other". Nowadays, it has the form $\Delta U = Q - W$ in which ΔU is the change in internal energy of the system, Q the heat released by the system and W the work done by the system (in the European convention, the firs law becomes $\Delta U = Q + W$). As we can see, this can also read as the law of conservation of energy, adapted to thermodynamic systems. The law of conservation of energy states that the total energy of an isolated system is constant: Energy can be transformed from one form into another, but it can neither be created nor destroyed. An important consequence of the first law is that perpetual motion (motion that continues indefinitely) cannot exist without an energy source!

In 1848, Thomson defines a new scale of temperature, which now bears his name: the Kelvin scale. This absolute scale aims to replace the old scales (Fahrenheit, Celsius, etc.), which are based on the behavior of a fluid (mercury, alcohol, water, etc.) in a tube, therefore limiting their use. Therefore, Lord Kelvin decides to use the efficiency of a Carnot engine as a measure of temperature. Since the calculation of the efficiency does not rely on the nature of the fuel, engine configuration or even its power, it is thus possible to think of a scale of absolute temperature that would be universal! He proposes an absolute temperature scale in which "a unit of heat descending from a body A at the temperature T of this scale, to a body B at the temperature (T-1), would give out the same mechanical effect, whatever be the number T". In other words, he proposes that the ratio of the heat received from the hot source to the heat yielded to the cold source be equal to the ratio of the absolute temperature of the hot source to that of the cold source; in other words, a Carnot engine, operating between 1,000K and 500K, rejects 500/1,000 = 50% of the heat it receives. If the low temperature is four times lower, the heat rejected is four times less (1,000/250) than the heat received. He also introduces the "absolute zero", corresponding to the temperature for which no more heat can be transferred. This theoretical temperature is now fixed, following an international agreement, at −273.15°C. Let us add that the temperature scale used nowadays is the Kelvin scale (K) and is not referred to as degrees. For example, 0K = −273.15°C. It is the base unit of temperature in the International System of Units (SI).

The second principle of thermodynamics starts from a historical statement; Thomson states "it is impossible, by means of an inanimate external agency, to derive mechanical effect from any portion of matter by cooling it below the temperature of the coldest of the surrounding objects". As we have seen before, according to Carnot, a thermal machine undergoes a caloric drop from the hot source to the cold source and it provides a mechanical work. However, the first principle tells us that there is energy conservation and work–heat equivalence. It is Clausius again, in 1856, who introduces another version of the second principle in his paper "On the moving force of heat": "Heat cannot, of itself, pass from a colder to a hotter body". A good example is the refrigerator, for which heat flows from cold to hot, but only when forced by an external agent, the refrigeration system!

Nevertheless, an issue remains. Machines, such as the Newcomen engine or other engines, are rather inefficient. Only 2% of the input energy is converted into a useful work output and a great amount of useful energy is dissipated or lost. To better understand this energy loss, Clausius introduces in 1865 the concept of entropy, S (Greek: τροπή = transformation). He uses the unit "Clausius" (Cl) for entropy. In nature, entropy is also revealed by the fact that heat always flows spontaneously from a hot body to a cold body. It never goes in reverse (cold → hot) spontaneously, except if work is done on the system. In other words, the total entropy (S) never decreases: It stays the same or increases! Mathematically, it reads as $\Delta S \geq \Delta Q/T$. At the end of Clausius' book "The Mechanical Theory of Heat: With Its Applications to the Steam-engine and to the Physical Properties of Bodies", we find the following two fundamental theorems of the mechanical theory of heat:

1. *"Die Energie der Welt ist constant"*. (The energy of the universe is constant.)
2. *"Die Entropie der Welt strebt einem Maximum zu"*. (The entropy of the universe tends to a maximum.)

There is also a third law of thermodynamics, often called Nernst's theorem. Indeed, Nernst (1864–1941) states that the entropy of a system at absolute zero is a well-defined constant. In 1912, he adds that "it is impossible for any procedure to lead to the isotherm T = 0 in a finite number of steps". An alternative version is given in 1923 by Lewis (1875–1946) and Randall (1888–1950): "If the entropy of each element in some (perfect) crystalline state be taken as zero at the absolute zero of temperature, every substance has a finite positive entropy; but at the absolute zero of temperature the entropy may become zero, and does so become in the case of perfect crystalline substances". In other words, S will reach 0 at 0K, as long as the crystal has a ground state with only one configuration, but we will see later why.

Finally, let us add that there is also a zeroth law of thermodynamics. It is first expressed by Maxwell in the words "All heat is of the same kind". Fowler (1889–1944) completes it in 1935: "If a body A is in temperature equilibrium with two bodies B and C, then B and C themselves will be in temperature equilibrium with each other". He also adds that "Any physical properties of A which change with the application of heat may be observed and utilized for the measurement of temperature". Perhaps more profound is the postulate from Fowler and Guggenheim (1901–1970): "we introduce the postulate: if two assemblies are each in thermal equilibrium with a third assembly, they are in thermal equilibrium with each other" and, later, "It may be shown to follow that the condition for thermal equilibrium between several assemblies is the equality of a certain single-valued function of the thermodynamic states of the assemblies, which may be called the temperature t, any one of the assemblies being used as a "thermometer" reading the temperature t on a suitable scale. This postulate of the "Existence of temperature" could with advantage be known as the zeroth law of thermodynamics."

THE CENTRAL ROLE OF ENTROPY

Let us start this section with another famous thought experiment designed in 1871 by Maxwell (1831–1879) and called "Maxwell's demon". Imagine a box filled with a gas at temperature T. This gas is composed of molecules moving at different velocities, i.e. following the Boltzmann–Maxwell distribution (sometimes at a velocity higher than the average speed, sometimes at a velocity lower than the average speed). Now, we add a partition, so that the box is divided into two compartments (left and right). The temperature is thus the same in the two parts. Right on the partition, there is a molecular-sized trap door with a demon at the doorstep. When a molecule has a velocity higher than the average speed, the demon opens the trap and lets the molecule go to the left compartment. On the other hand, when the molecule has a velocity lower than the average speed, he opens the trap and lets the molecule go to the right compartment. Therefore, the box becomes hot on the left and cold on the right! This means that one can use this difference in temperature to run a heat engine, by allowing the heat to flow from the hot side to the cold side. Doing so, it is possible to create a perpetual motion machine! Also, let us note that the temperature on the left has increased while the temperature on the right has decreased, meaning that the right part of the box has cooled down thanks to the hot part of the box! Yet, according to the second law of thermodynamics, the heat transfer should take place from a hot body to a cold body and not the inverse. Moreover, having a box with a partition decreases the total entropy of the box, because the volume of each compartment is lower than the total volume of the box. Another important point is that, since the demon makes a choice based on the velocities of the molecules for each compartment, he ends up creating some "order" in each compartment, again decreasing the total entropy of the box. The Maxwell's demon experiment highlights a paradox: There would be a way to violate the second law of thermodynamics! This is a tremendously stimulating and intriguing problem that many scientists work on for more than a half-century!

The development of statistics provides a new means to understand thermodynamics. As we have seen, the latter deals with macroscopic quantities such as temperature, volume and pressure. Statistics, on the other hand, is interested in the behavior of each molecule, or in other words, in microscopic quantities such as molecular velocity and molecular position. However, in practice, it is impossible to follow each molecule. For example, one liter of air contains about 3×10^{22} molecules (of the order of Avogadro's number)! There is therefore a need to use statistics as with the Maxwell–Boltzmann distribution for the velocities of molecules in a gas at a given temperature!

Boltzmann (1844–1906) formulates an equation for the entropy between 1872 and 1875: the famous $S = k_B \ln W$. According to him, the entropy of an ideal gas is related to probability through the quantity W (German: *Wahrscheinlichkeit* = probability). W represents the number of microscopic arrangements (configurations) corresponding to the gas macroscopic property entropy S and k_B the Boltzmann constant. This equation links statistics and thermodynamics through the definition of thermodynamic entropy. For the anecdote, it is Planck, later, who writes the current form of the formula. Boltzmann, working with gram molecules and not in terms of molecules themselves, did not include the factor before the logarithm, the well-known k_B!

Gibbs (1839–1903), who happened to be the first person to receive a Doctorate of Engineering in the USA, creates statistical mechanics with Maxwell and Boltzmann. Let us add that Gibbs publishes his first major article in 1873, "Graphical Methods in the Thermodynamics of Fluids", in which he presents the geometric representation of thermodynamic state functions (see Figure 2.15). Later, in his now famous

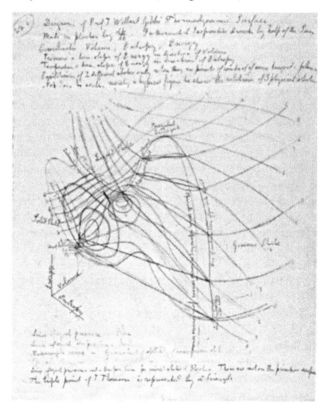

FIGURE 2.15 Maxwell's sketch of thermodynamic surfaces, based on the theory proposed by Gibbs (1875).

"On the Equilibrium of Heterogeneous Substances", he defines the "heat function for constant pressure", that we know now as enthalpy, the "potential" now chemical potential and as well as the "available energy". The latter is now named Gibbs free energy in his honor. He also creates statistical sets, now known as the canonical ensemble, microcanonical ensemble and grand-canonical ensemble. Because of his interest in statistics, he introduces the fact that the microstates of a thermodynamic system are not equally probable, leading to a new definition of entropy in 1878 with $S = -k_B\sum p_i \ln p_i$

Let us add that he is also known for the famous Gibbs' paradox. This is another thought experiment in which a box is divided into two compartments. There are two scenarios. First, we have the same gas in both compartments. In that case, when the partition is open, there is mixing of the molecules and the thermodynamic entropy stays the same. The same is true when we put the partition back in place. Second, we have two different gases in each compartment. Once we open the partition, both gases mix, and we end up with a thermodynamic entropy different from the two initial entropies of each compartment. By doing the calculation, we find an entropy increase of $2Nk_B\ln 2$ which, as we can see, does not depend on the nature of the gas but only on the number of molecules N and the fact that they are different (ln 2): "the increase of entropy due to the mixing of given volumes of the gases at a given temperature and pressure would be independent of the degree of similarity between them". The paradox here is the following: As the two gases become more and more similar, the entropy of mixing does not smoothly go to zero, but it suddenly drops to zero in a discontinuous way! The difference in identity between two molecules will play a major role in quantum mechanics with the existence of the concept of indistinguishability.

In 1948, a new advance is made by Shannon (1916–2001). Indeed, he publishes "A Mathematical Theory of Communication" and introduces the concept of information entropy. A real revolution! He proposes a new measure of the efficiency called entropy but, this time, for communications systems in bits (or "Shannons"). The idea is to detect repeating patterns in information and optimize communication by getting rid of the superfluous content. In practice, this is done using a lossless compression algorithm, that removes information redundancy by locating repeated patterns. For example, the sentence "Thank you for your attention" has a Shannon entropy of 4.75. To represent this 24-character string, $24 \times 4.75 = 114$ bits are required if the string is encoded optimally! Using a simple Shannon–Fano coding, one obtains the following sequence:

0111111011101101011111110111101011111011111111101011111101111010111110 11111101110001111111111011001111111111110110.

This message can be decoded using the following dictionary: T:0, O:10, N:110, A:1110, Y:11110, U:111110, R:1111110, H:11111110, K:111111110, F:1111111110, E:11111111110, I:111111111110. Here "T" is coded with the fewest bits (a single 0), since it is the letter most often used in the sentence. This reasoning is nowadays very frequently used in data compression such as ZIP files, MP3s and JPEGs. It also allows for the development of many applications in cryptography, cryptocurrency (bitcoins), quantum computing, language coding and screening tests in medicine.

   Using this new information theory, Brillouin (1889–1969) makes a breakthrough on the Maxwell's demon paradox. He demonstrates that the decrease in entropy, arising from the demon's actions, is less than the entropy increase resulting from making a choice between the fast and slow molecules. In an article published in 1951, he discusses that, unless a source of light is introduced in the box, the demon cannot see the molecules and operate the trap door. Indeed, the demon cannot see the individual molecules since, in an enclosure at constant temperature, the radiation is that of a "blackbody". Just recall that it is only during the 20th century that Planck develops quantum theory, giving an answer to the blackbody mystery. Once the source of light is accounted for, the overall balance of entropy becomes positive again. He summarizes his finding with the following cycle: negentropy → information → negentropy (here negentropy stands for negative entropy). Moreover, he shows that the Boltzmann constant (k_B) is the smallest amount of negative entropy required in an observation. This saves the second law of thermodynamics!

   One of Brillouin's students, Landauer (1927–1999), is interested in understanding the limitations associated with building a machine capable of carrying out computations. In 1961, he proposes the famous Landauer's principle: "any logically irreversible manipulation of information, such as the erasure of a bit or the merging of two computation paths, must be accompanied by a corresponding entropy increase in non-information bearing degrees of freedom of the information processing apparatus or its environment". In particular, he shows that when information is lost, there is a release of a quantity of $k_B T \ln 2$ of heat!

   Let us add that in 1944, Schrödinger introduces the negative entropy in his book "What Is Life?: The Physical Aspect of the Living Cell". He states that: "The orderliness encountered in the unfolding of life springs from a different source. It appears that there are two different mechanisms by which orderly events can be produced: the statistical mechanism which produces order from disorder and the new one producing order from order." Regarding Nature, he writes that: "Thus a living organism continually increases its entropy – or, as you may say, produces positive entropy – and thus tends to approach the dangerous state of maximum entropy, which is of death. It can only keep aloof from it, i.e. alive, by continually drawing from its environment negative entropy." Finally, he adds: "You would not expect two entirely different mechanisms to bring about the same type of law – you would not expect your latch-key to open your neighbor's door as well".

BEYOND EQUILIBRIUM: THE RISE OF NONEQUILIBRIUM THERMODYNAMICS

Perhaps another interesting question Schrödinger asks in his book is "How does the living organism avoid decay? [...] The obvious answer is: by eating, drinking, breathing and (in the case of plants) assimilating. The technical term is metabolism". Indeed, he posits that living organisms stay alive by fending off equilibrium. Or in other words, remaining alive is a constant fight against disorder. For instance, humans release heat to the surroundings, yet steer clear of running out of energy by regularly ingesting food! At the same time, we know from the second law of thermodynamics that the entropy of a system always increases. As entropy increases, the system becomes more and more disordered. Since $S = k_B \ln W$, it also implies that the probability W becomes greater, leading to an increased complexity of the system! However, Nature teaches us that living organisms choose complexity, but also highly ordered structures characterized by a low entropy (school of fish, flocks of birds) to survive. To reconcile these two facts, a conclusion must be drawn: Nature operates far from equilibrium!

On the theoretical front, one of the most important advances comes from Onsager (1903–1976). In 1968, he is awarded the Nobel Prize in Chemistry "for the discovery of the reciprocal relations bearing his name, which are fundamental for the thermodynamics of irreversible processes". Irreversible processes, which are the most common in nature, present the characteristic of not being able to go backward. For example, if we pour some milk in a cup of tea, the two liquids mix. There is no way the mixture is going to separate spontaneously at any point in the future. In other words, we cannot have the cup of tea without milk and the milk back in its bottle! This is where Onsager starts his discoveries. Indeed, he proposes what it is often called the "fourth law of thermodynamics", which deals with irreversible processes. The idea is the following: If we take our example of the tea and the milk, diffusion (mixing tea and milk) and heat transfer (hot tea and cold milk) occur simultaneously! Indeed, both phenomena impact one another. Diffusion, in addition to inducing a flow in matter, also triggers a heat flow. Conversely, temperature differences, in addition to activating a heat flow, also result in a flow of matter. There is thus a connection between irreversibility and complexity! Thanks to the Onsager reciprocal relations, connections can be made between equations governing different flows, providing a full picture of irreversible processes!

Yet the spontaneous organization of water during convection is mind-boggling. Consider a thin layer of water and begin heating it from the bottom. After some time, the heat reaches the top of the layer and water starts to form highly organized, honeycomb-like structures called Bénard cells. As heat is supplied, the system is driven out of equilibrium. Heat is then used by the system to do "work" (life) and create highly ordered structures. When the heat supply stops, the system goes back to equilibrium and the highly organized structures disappear (death).

A first explanation of the onset of these highly ordered structures comes from Prigogine (1917–2003) (see Figure 2.16), who receives the Nobel Prize in Chemistry in 1977 "for his contributions to non-equilibrium thermodynamics, particularly the theory of dissipative structures". Indeed, he shows that the "Bénard instability" can only be created if the temperature difference between the top and the bottom of the layer is large enough. Since this phenomenon is far from equilibrium, there will be entropy production, as long as energy is supplied. Here, in addition to the heat flow, there is also a flow of matter, which means that the entropy production is increased! More and more molecules move in a concerted way over time and start to organize. They finally achieve the formation of the "honeycomb" pattern, demonstrating that nonequilibrium can be a "source of order"!

He also finds that "dissipative structures" exist not only in the case of flows but also for chemical reactions. To do so, he devises a model for autocatalytic reactions to analyze instabilities in oscillating chemical systems. This model is called the "Brusselator" (a contraction of Brussels, capital of Belgium where Prigogine

FIGURE 2.16   Ilya Prigogine (1917–2003).

works, and of the word oscillator). In particular, he chooses a specific frequency of oscillations that allows him to realize a chemical clock! In mathematics, this is called a Hopf bifurcation, meaning that the system starts to exhibit a periodic (oscillatory) behavior. Creating these instabilities lead to the formation of dissipative structures. In particular, he observes that "there are three aspects which are always linked in dissipative structures: the function as expressed by the chemical equations, the space-time structure, which results from the instabilities, and the fluctuations, which trigger the instabilities. The interplay between these three aspects leads to most unexpected phenomena, including 'order through fluctuations'." According to Prigogine, this provides the ultimate proof that "the arrow of time is not an illusion created by us".

Let us add that an especially striking example can be found in the famous Belousov–Zhabotinsky reaction (BZ reaction) in chemistry. It is also an autocatalytic chemical reaction that exhibits an oscillatory behavior. In the early 1950s, Belousov thinks of reproducing the human metabolism in a test tube. He devises a set of chemical reactions mimicking the Krebs cycle, which converts organic molecules into carbon dioxide and water. To do so, he first uses citric acid (organic molecule), bromate ions (oxidizing agent, playing the role of stimulus) and cerium ions (transition metal, serving as catalyst). He mixes the three reactants and obtains carbon dioxide and water. What is especially impressive is that the color of the test tube changes color periodically, as the reaction proceeds. In a matter of seconds, the solution turns from colorless to yellow, and back again to colorless, then yellow and so on! Zhabotinski continues Belousov's work and finds that using ferroin as a catalyst and malonic acid as the organic molecule gives rise to the formation of patterns! Indeed, he observes that oxidation waves propagate in the reactor, leading to the formation of blue oxidation fronts on a red background (reduced ferroin). Every

FIGURE 2.17   Alan Turing (1912–1954).

time a wave is broken, this creates a pair of spiral waves, giving rise to an entire family of new ordered structures. In fact, the theory of chemical oscillators extends far beyond chemistry and is crucial for many biological systems. For example, Turing (1912–1954), the very one from the movie "The Imitation Game" (see Figure 2.17), studies pattern formation through chemical instabilities during biological morphogenesis (for instance, the formation of stripes in zebras). Another example can be found in the behavior of the *Dictyostelium discoideum*, a mold that forms spiral patterns similar to those exhibited by the BZ reaction.

Furthermore, a compelling difference between chemical and biological systems is the possibility, for biological systems, to use the coded information contained in their DNA (or RNA for viruses) to respond to changes in their environment. This coded information was foreseen by Schrödinger in 1944: "Organic chemistry, indeed in investigating more and more complicated molecules, has come very much nearer to that 'aperiodic crystal' which, in my opinion, is the material carrier of life".

Let us add that Watson, Crick and Wilkins, Nobel laureates in Physiology or Medicine in 1962 "for their discoveries concerning the molecular structure of nucleic acids and its significance for information transfer in living material", have all written that they were inspired to study DNA after reading "What Is Life?" by Schrödinger.

3 Introduction to Quantum Mechanics
Wave–Particle Duality

On en dirait autant de la perception: auxiliaire de l'action, elle isole, dans l'ensemble de la réalité, ce qui nous intéresse; elle nous montre moins les choses mêmes que le parti que nous en pouvons tirer. Par avance elle les classe, par avance elle les étiquette; nous regardons à peine l'objet, il nous suffit de savoir à quelle catégorie il appartient. Mais, de loin en loin, par un accident heureux, naissent des hommes qui, soit par leurs sens soit par leur conscience sont moins attachés à la vie. La nature a oublié d'attacher leur faculté de percevoir à leur faculté d'agir. Quand ils regardent une chose, ils la voient pour elle, et non plus pour eux. Ils ne perçoivent plus simplement en vue d'agir; ils perçoivent pour percevoir - pour rien, pour le plaisir.

Bergson, "*La perception du changement*", Conferences, 1911

One could say as much for perception: the auxiliary of action, it isolates that part of reality as a whole that interests us; it shows us less the things themselves than the use we can make of them. It classifies, it labels them beforehand; we scarcely look at the object, it is enough for us to know to which category it belongs. But now and then, by a lucky accident, men arise whose senses or whose consciousness are less adherent to life. Nature has forgotten to attach their faculty of perceiving to their faculty of acting. When they look at a thing, they see it for itself, and not for themselves. They do not perceive simply with a view to action; they perceive in order to perceive, - for nothing, for the pleasure of doing so.

Bergson, "The Philosophy of Change", Conferences, 1911

BIRTH OF A PARTICLE THEORY OF LIGHT

Until the 18th century, very little is understood about light. According to the philosophers Euclid and Plato, light is produced by our eyes. Rays of light are beams from the eyes that allow us to see the objects surrounding us, just like the headlights of a car. This is the reason why ancient Greeks and Romans use glass or stone insets to represent the eyes in their sculptures! Even if their description of vision seems quite off to us, they are already aware of one major law: the law of reflection. First, Euclid states that light travels in a straight line. Second, he notices that a ray of light striking a mirror with a specific angle will bounce back, or reflect, from this mirror with the same angle! In other words, when we observe an object in a mirror, we actually look at an image of this object, which is the same as the object but reversed. It is captured by the phrase "mirror image"! This is what we see, for example, when a mountain is reflected in a lake. In this case, the lake plays the role of a mirror and the mountain appears upside down since it is reversed! Furthermore, a molecule that cannot be superimposed on its mirror image is called a chiral molecule, and this is pretty much the case for many molecules in biochemistry including sugars!

Yet another phenomenon remains difficult to understand. For example, let us dip a spoon in a water bowl. If one looks under the surface of water, the spoon appears closer than it really is! Moreover, if one looks at the straight part of the spoon, it appears to bend at the water surface! Other intriguing phenomena such as rainbows and mirages cannot be explained with the law of reflection. Are these illusions? What about the seemingly "flattened sun" when the sun gets close to the horizon? The beginning of an answer comes from Snell (1580–1626). As a mathematician, he is passionate about finding a value for π using polygons. The first scientist to use the polygon method is van Ceulen (1540–1610). He finds, in 1596, a value of π accurate to its first 20 places! Using a polygon of 2^{62} sides (~5,000,000,000,000,000,000 sides), Snell succeeds in calculating π in 1621 as 3.1415926535897932384626433832 7950288 $< \pi <$ 3.1415926535897932384626433327950289. In other words, his value for π is exact to 35 places! He also publishes "sine tables". After some calculations, he manages to find a new mathematical formula that explains the famous "bending" of objects at the air–water interface. Indeed, he relates the angle of the ray of light that travels through air (θ_{air}) to the angle of the ray when it goes through water (θ_{water}). He notices that the ratio of the sine of the first angle to the sine of the second angle is constant. In particular, this number is equal to 1.33 for water. We call it today the refractive index (n). Mathematically, this law is called Snell's law and reads as $n_{air} \sin \theta_{air} = n_{water} \sin \theta_{water}$. Because of refraction, the ray of light changes its path at the interface (see Figure 3.1). This is the reason why, for instance, the spoon appears to bend, since our eyes actually see the two different paths of light in air and in water.

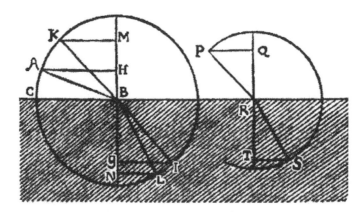

FIGURE 3.1   Refraction laws from Descartes' *"Dioptrique"* (1637).

Since light follows Fermat's principle and travels along the quickest path, the change in path experienced by light also means that the velocity of light is not the same in air and in water. This is exactly what the refraction index measures, namely that light travels 1.33 times more slowly in water than in air!

Let us add that refraction (or refractometry) is also the name of a clinical test during an eye exam. The idea is to determine if the eye has a refractive error, and if so, what best corrective lenses should be prescribed. In practice, several test lenses with different optical powers are presented to the patient. The one giving the sharpest, clearest, vision is then selected, and the refractive error can be determined. For example, an eye that has no refraction error when viewing a distant object 6 meters or 20 feet away is said to have emmetropia (Greek: *en-* (in) + *metron* (measure) + *-ops* (eye)). It means that the eye can focus parallel rays of light in the retina without using any accommodation. Therefore, one will score "6/6" (meters) or "20/20" (feet). For example, a person with "20/80" vision can only see clearly an object at 20 feet whereas others who do not need corrective lenses can see at 80 feet, meaning that this person is nearsighted or has myopia. The purpose of the corrective lenses is to compensate for refractive errors of the eyes including myopia, hyperopia and astigmatism (see Figure 3.2). The concept of corrective lenses is to help focus the ray of light on the retina, thus allowing a person to have a clear image of the object. More specifically, for an eye exhibiting myopia, the eye focuses the ray of light before it actually reaches the retina, giving a blurred image. Therefore, the corrective lens should be concave in order to make the ray diverge, allowing the eye to focus exactly on the retina. For hyperopia, the eye focuses beyond the retina surface. To compensate for that, the corrective lens should be convex in order to make the ray converge, again helping the eye to focus on the retina. In practice, the corrective lenses are characterized by their optical power measured in diopter

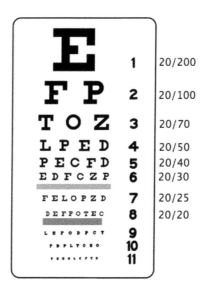

FIGURE 3.2   Snellen chart.

units (D) with $1 D = 1 m^{-1}$. For example, $-2.0 D$ corresponds to a power for the corrective lens of $2 m^{-1}$ meaning that it creates a focal point at 0.5 m or 50 cm. The negative sign indicates that the focal point is located 50 cm before the eye, implying that this prescription is intended to correct myopia. On the other hand, there is no negative sign in the case of corrective lenses for hyperopia, since the focal point should be after the eye. But the story does not end there. How about distance and depth perception? 3D vision is achieved through our ability to have binocular vision (two eyes). Both eyes help us see an image from two different points and angles. Then, the brain processes the information from optic nerves and uses different techniques to assess distance and depth perception. For example, convergence and accommodation are two different techniques the brain uses to estimate distances, based on efforts made by eye muscles. For instance, an object close to us requires more effort than the accommodation needed to see an object further away. More information can be obtained through parallax and geometry pattern recognition, which we acquire as we grow up. This gigantic database is then used by the brain to apprehend the world surrounding us, in 3D. Let us not forget that 33% of our cerebral cortex plays a role in processing vision!

Incidentally, the name diopter comes from Kepler and his book *"Dioptrice"* in which he discusses lenses and the making of telescopes. In *"La Dioptrique"* (1637), Descartes studies the properties of light. According to him, light can be seen as a baton that allows one to see by night, clarifying ancient Greeks' ideas. He uses the image of a baton to describe the linear nature of a ray of light. Moreover, he designs an experiment with a ball traveling along a line to try to explain clearly the laws of reflection and refraction. The ball motion can be damped by soft bodies (refraction) or it can bounce back on hard bodies (reflection). Descartes obtains the same laws as Snell, which is why the laws of reflection and refraction are now known under the name of laws of Snell–Descartes. However, his theory predicts that the light goes faster in water than in air! Let us add that Descartes also paves the way for the invention of the first pair of glasses. *"Par la dioptrique, j'eus le dessein de faire valoir qu'on pouvait aller fort avant dans la philosophie pour arriver par son moyen jusqu'à la connaissance des arts qui sont utiles à la vie, à cause que l'invention des lunettes d'approches, que j'y expliquais, est l'une des plus difficiles qui aient jamais été traitées"*.

In 1704, Newton publishes "Opticks: or, A Treatise of the Reflexions, Refractions, Inflexions and Colours of Light" that gathers all his experiments and theories on light. It remains a reference book until the 19th century! In his book, he presents the first explanation of the decomposition of light by a prism (dispersion). For this purpose, he directs sun rays through a prism. This produces a rainbow! At this time, it is thought that the glass of the prism has a hidden color that is revealed as the sun rays pass through. He then shows that if this rainbow passes through a second prism, there is, again, a white light! His conclusion is revolutionary: the color is in the light and not in the glass! He thus deduces a theory of colors: The white light that we see is, in fact, a mixture of all the colors in the spectrum visible by the eye! He also introduces in 1672 his "corpuscular" or particle theory of light. Indeed, Newton explains that light is made up of small discrete particles called "corpuscles" (little particles), which travel along a straight line with a finite velocity and possess "impetus" (energy from their motion). The rays of light are then the trajectory of the corpuscles. The reflection corresponds to their bouncing back on a mirror. Refraction is explained by the fact that corpuscles of light interact with matter. Colors can also be interpreted as many varieties of corpuscles. Note also that Newton discovers another interesting phenomenon, now known as Newton's rings, in 1717. When a convex lens is placed on a flat surface, there is a series of concentric rings, alternatively bright and dark, centered on the point of contact. Despite all his efforts, Newton never manages to explain this phenomenon, which will only be understood later under the name of interferences!

NEW DEVELOPMENTS PROMPT A WAVE THEORY OF LIGHT

One of the most surprising experiments is performed by Grimaldi (1618–1663). After piercing a pinhole in the curtains, Grimaldi observes that rays of light produce a luminous ellipse on a screen. He then places different objects after the pinhole and finds very surprising images. For instance, a rod gives a shadow that is much larger than the actual size of the rod and, on the edges of the image, there are colored bands. If one replaces the rod by a bird's feather, an even stranger image appears. The shape of the quill looks rather like irrigated fringes, and several points on the screen, which are not in the line of the pinhole, are lit nonetheless. Yet, the light can only go through the small interstices separating the barbs of the feather, and there is no diopter in this experiment that could account for any deviation of light. He names these phenomena as "diffraction" of light, from the Latin word *diffractio* (break apart) (see Figure 3.3).

To account for these strange observations, Huygens proposes a new theory explaining the nature of light. In 1690, he publishes his famous "*Traité de la*

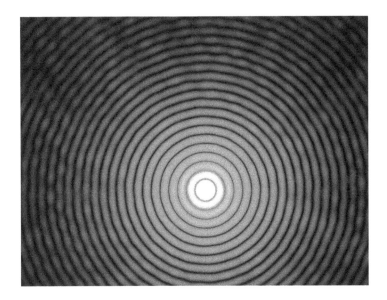

FIGURE 3.3 Diffraction patterns.

Lumière" (Treatise on Light). His idea is that light is composed of waves, in total opposition to Newton's theory! It starts with, "As happens in all the sciences in which Geometry is applied to matter, the demonstrations concerning Optics are founded on truths drawn from experience. Such are that the rays of light are propagated in straight lines; that the angles of reflexion and of incidence are equal; and that in refraction the ray is bent according to the law of sines." According to him, light travels like sound. He states, "We know that by means of the air, which is an invisible and impalpable body, Sound spreads around the spot where it has been produced, by a movement which is passed on successively from one part of the air to another; and that the spreading of this movement, taking place equally rapidly on all sides, ought to form spherical surfaces ever enlarging and which strike our ears." He adds that light "spreads, as sound does, by spherical surfaces and waves: for I call them waves from their resemblance to those which are seen to be formed in water when a stone is thrown into it, and which present successive spreading circles." In other words, light radiates wavefronts! For example, the wavefront created by a stone thrown in a lake has a perimeter of $2\pi R$. For sound, which is a wave in 3D, the wavefront is the area of a sphere: $4\pi R^2$. Huygens also defines "ethereal matter", the matter in which the movement coming from a luminous body is propagated. It is not the same medium as for sound, since, unlike sound, light continues its path when air is removed from a vessel. According to his calculation on an eclipsed Jupiter's moon, he finds that the speed of light is much greater than the speed of sound! Nowadays, we now know that the speed of sound at 20°C is about 3×10^2 m/s, whereas the speed of light is 3×10^8 m/s, giving us a ratio of 1×10^6, meaning that light travels a million times faster than sound! He adds that, to propagate, light spreads successively as spherical waves: "the propagation consists nowise in the transport of those particles but merely in a small agitation which they cannot help communicating to those surrounding, notwithstanding any movement which may act on them causing them to be changing positions amongst themselves". He continues, "this prodigious quantity of waves which traverse one another without confusion and without effacing one another must not be deemed inconceivable; it being certain that one and the same particle of matter can serve for many waves coming from different sides or even from contrary direction." Thanks to this theory, he succeeds in deriving the laws of reflection and refraction, as well as diffraction effects. The issue is that the phenomenon of refraction shows that light travels faster in air than in dense media. Unlike particles, waves propagate faster in dense media than in air (see the example of sound). In addition, light is clearly capable of traveling through an interstellar void, while a wave needs a medium for material propagation to exist. However, because of Newton's reputation, the corpuscular theory persists until the 19th century.

In 1801, Young (1773–1829) carries out a very important experiment. It is now called Young's interference experiment or Young's double-slit interferometer (see Figure 3.4). Young's experiment is very similar to Grimaldi's, except that he uses thin vertical slits. First, he studies the pattern obtained with a single slit. When he shines a light through the slit, he observes a series of alternating dark and bright fringes on the screen. Redoing the experiments with two slits, he notes that some fringes that were previously bright have now become dark, and conversely. In short,

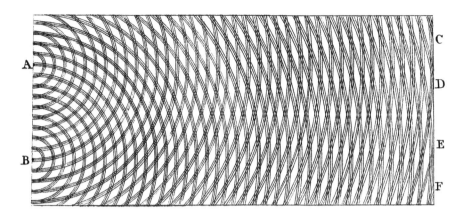

FIGURE 3.4   Young's double-slit interferometer.

by adding more light with two slits instead of one, he manages to darken fringes that were previously illuminated! Young therefore hypothesizes that light behaves as a sinusoidal wave: "But the general law, by which all these appearances are governed, may be very easily deduced from the interference of two coincident undulations, which either cooperate, or destroy each other, in the same manner as two musical notes produce an alternate intention and remission, in the beating of an imperfect unison". He shows that, thanks to the wave theory, it is now possible to understand these phenomena of interferences: Constructive interferences of two waves lead to a bright stripe, whereas destructive interferences of two waves result in a dark stripe! He also demonstrates that wave theory can account for the existence of different colors. Indeed, each color has a characteristic frequency (υ) and wavelength (λ)! "Supposing the light of any given colour to consist of undulations of a given breadth, or of a given frequency, it follows that these undulations must be liable to those effects which we have already examined in the case of the waves of water and the pulses of sound [...] We are now to apply the same principles to the alternate union and extinction of colours." Young is even capable of determining a strikingly accurate wavelength for different colors. In particular, he finds $\lambda = 576$ nm for the color yellow (within 1 nm of the modern estimate!).

Incidentally, the French Academy of Sciences sets the subject of the 1818 prize of the Academy as "diffraction". At that time, the corpuscular theory is still dominant among scientists. Fresnel (1788–1827) submits a paper based on wave theory combining Huygens' principle and Young's interference work. Fresnel's work is rejected by the committee, and especially by Poisson who considers that this paper presents absurd conclusions. It will become the famous "Poisson spot" or "Fresnel bright spot" problem. Indeed, according to Fresnel, there should be a bright spot in the middle of a shadow on a circular screen, while, according to Poisson, there should no such bright spot. Arago (head of the committee) decides to perform the experiment to find out who is right. To everyone's surprise, Arago observes a bright spot, convincing most scientists of the wave nature of light! In the end, Fresnel wins the competition. In 1821, Fresnel uses integral calculus to predict from wave theory all kinds of diffraction and interference patterns!

Another question remains: what is this medium called "ether" in which light waves are supposed to propagate? To answer this question, Michelson and Morley develop their interferometer in 1887. Their idea is that, if light is a wave, it therefore needs a medium to propagate. At that time, the Universe is immersed in ether. Michelson and Morley design the interferometer in the hope to detect this famous ether. However, it does not detect anything! For years, they conduct again, again and again, their measurements, regularly improving their device. In 1887, they must come to the following conclusion: ether does not exist! Thus, the so-called "light waves" do not have a medium to propagate in, meaning that they cannot exist either! Eventually betrayed by an interferometer, the wave theory turns out to be a colossus with feet of clay. Deprived of one of its fundamental assumptions, this theory collapses like a house of cards.

THE DAWN OF SPECTROSCOPY

Nevertheless, a few experiments on the solar spectrum will change the course of history. Indeed, Newton's experiments using prisms open a new way to look at light refraction. The astronomer Herschel (1738–1822) works with his sister Caroline (also a well-known astronomer) to build large telescopes and identify celestial objects such as stars, moons and comets. In 1781, he finds a new object near the Gemini constellation: the planet Uranus! Following this discovery, George III (King of England) becomes his patron and gives him access to more and more powerful telescopes, allowing for the discovery of Titania and Oberon (moons of Uranus) and Enceladus and Mimas (moons of Saturn)! Mimas is also known to astronomers as "Saturn's Death Star moon" as it closely resembles the famous "Death Star" from the *Star Wars* movies. In 1800, Herschel uses a prism and a thermometer to study sunlight and measure the temperature of each color he obtains. He finds that there is a specific temperature corresponding to a given color! He also notes that, when he determines the temperature for the color red, it is the highest of the entire spectrum. Another of his observations is quite astonishing. When he places the thermometer after the color red (where there is no more color), the temperature rises! It means that there is something else there, that has the properties of light but is not visible to the eye. He names it "heat rays", which we now call infrared light or infrared radiation (Latin: *infra* = below). Nowadays, infrared technology is used, for example, in night-vision devices to observe living beings by night. These apparatuses read body-temperature thermal light or, in other words, the infrared radiations they emit.

Ritter (1776–1810) is fascinated by this new discovery. According to him, if there is an infrared light, there should also be an ultraviolet radiation at the other end of the visible spectrum. After a series of experiments using different chemical compounds, he notices that stripes of silver chloride change faster from white to black when they are placed in the dark region of the Sun's spectrum (after its violet end). There is thus another light after the violet ray that Ritter calls "chemical rays". They are now named "ultraviolet radiations" (Latin: *ultra* = beyond) as the counterpoint to infrared radiation! Let us add that experiments with silver chloride, silver bromide and silver iodide (known as the family of silver halides), will have a tremendous impact during the 19th century. Indeed, these will be at the center of photography! The idea behind this invention is that, when exposed to light, silver halides produce metallic silver that darkens the film to create a black and white picture! In 1842, Antoine Becquerel (1788–1878) is the first to photograph the solar spectrum using the discovery of Daguerre (photography). He also discovers, with his son Edmond, that illuminating certain materials causes the passage of an electric current. Indeed, in his electrochemical experiment, silver chloride or silver bromide are used to coat platinum electrodes; once the electrodes are illuminated, voltage and current are generated, leading to the first photovoltaic cell! Let us add that the Becquerel family produces four generations of great physicists. Antoine's grandson Henri receives the Nobel Prize in Physics in 1903 for the discovery of radioactivity alongside Pierre and Marie Curie!

The development of spectroscopy as we know it now really starts with Fraunhofer (1787–1826). Indeed, in 1814, he designs his spectroscope with a prism, a diffraction slit and a telescope. During his experiments, he remarks that the light emitted by a fire appears, through his spectroscope, in the orange color of the spectrum. He then decides to see if the solar spectrum also contains the same orange bright line. Thanks to his new apparatus, he rediscovers what Wollaston has previously seen; there are many dark features in the solar spectrum! He then systematically studies and measures where these features are. "In looking at this spectrum for the bright line, which I had discovered in a spectrum of artificial light, I discovered instead an infinite number of vertical lines of different thicknesses. These lines are darker than the rest of the spectrum, and some of them appear entirely black." In all, he maps over 574 dark lines, designating the principal features (lines) with letters A through K (see Figure 3.5). They are now known as Fraunhofer lines! Nowadays, modern observations of sunlight can detect over 25,000 dark lines. So, a new question arises: What do these black lines correspond to?

In 1826, Talbot (1800–1877) publishes "Some Experiments on Colored Flames" and observes the flames produced by salts of sodium (Na) and potassium (K). "This red ray appears [...] to be characteristic of the salts of potash, as the yellow ray is of the salts of soda." From 1834, in a series of papers entitled "Facts Relating to Optical Science", Talbot notes that the spectral flames of strontium (Sr) and lithium (Li) look indistinguishable. However, he writes that the strontium flame shows many red lines separated by dark intervals, whereas lithium only exhibits a single red line! This is the beginning of spectrochemical analysis! He

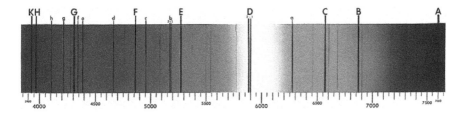

FIGURE 3.5   Fraunhofer lines.

also writes about the dark lines in the spectrum produced by iodine vapor: "I have found by careful observation that they are not equidistant, but that they become gradually more crowded towards the blue end of the spectrum. This [...] seems a consequence of some simple general law. [...] I have advanced the hypothesis that the vibrations of light and those of material molecules are capable of mutually influencing each other. It remains to be seen, whether the same hypothesis does not afford a clue to the explanation of this apparently complex phenomenon of absorption."

Around 1859, Kirchhoff (1824–1887) and Bunsen (1811–1899) find a definitive answer. Thanks to Bunsen's famous burner, they manage to study line spectra with a great sharpness and discover that the dark lines are indeed atomic absorption lines! Indeed, when a white light passes through a chemical substance, and then through a prism, an absorption spectrum appears! In other words, the chemical substance "absorbs" radiations corresponding to the dark lines! Emission is the inverse process and therefore produces bright lines. In fact, when the light emitted by an element passes through a prism, a bright line spectrum is produced! Atomic emission spectra are characteristic of an element and, as such, are often called chemical fingerprints (see Figure 3.6). They also relate absorption and emission through the so-called Kirchhoff laws. The first law states that a hot solid, liquid or dense gas emits a continuous spectrum of radiation. This means that, if the light emitted passes through a prism, there will be a continuous band showing all colors of the rainbow! The second law states that a dilute hot gas emits bright lines called "emission lines" of the spectrum. In this case, the radiation passing through a prism would yield a series of isolated lines of different colors. The third law states that, when a white light passes through a cool dense gas, it produces dark lines or "absorption lines". Approximately 50 years after Fraunhofer's discovery, Kirchhoff and Bunsen notice that several Fraunhofer lines match characteristic emission lines identified in the spectra of heated elements. They thus conclude that the dark lines in the solar spectrum are due to the absorption of light by chemical elements around the sun! In 1860, they publish *"Chemische Analyse durch Spectralbeobatchtungen"*, in which they describe their spectroscope and apply it to discover cesium (Cs) and rubidium (Rb) through the analysis of their emission spectra. Let us also add that Kirchhoff introduces in 1862 the concept of a perfect blackbody that represents an idealized physical body absorbing all incident radiation, regardless of frequency or angle of incidence.

Incidentally, an unknown chemical element is identified in the spectrum of the sun in 1868 by many astronomers during the eclipse of the sun. Lockyer in 1868 determines the wavelength of this ray at 5,876 Å. This element was thus named "helium" (Greek: *helios* = sun). This gas is only identified on Earth in 1895!

FIGURE 3.6   Emission spectrum for hydrogen.

BLACKBODY RADIATION AND ULTRAVIOLET CATASTROPHE

In 1862, a *coup de théâtre*: Maxwell (1831–1879) develops mathematical equations that predict the existence of waves traveling in vacuum; he calls them "electromagnetic waves" and even deduces their speed of propagation! "This velocity is so nearly that of light that it seems we have strong reason to conclude that light itself (including radiant heat and other radiations) is an electromagnetic disturbance in the form of waves". However, this is just a theory and it needs to be verified. The first to generate and detect an electromagnetic wave is Hertz (1857–1894). He carries out several experiments using "Riess or Knochenhauer spirals" and a "Leyden jar", reported in 1893 in his book "Electric Waves". A Leyden jar is a glass bottle filled with water and with a wire passing through its cork. To charge the Leyden jar, the wire is put in contact with an object that has been previously rubbed, leading to a jar filled with electricity. As an aside, if a group of people holds hands, and one of them touches the wire of a charged Leyden jar, everybody, one after the other, will feel a small electric shock passing through, also known as the famous "ring of fire" experiment. Nowadays, a Leyden jar is known as a capacitor or condenser. As for a Riess spiral (named after the German physicist Riess), it is made of brass knobs, connected by a spirally wound wire. Nowadays, it is known as an induction coil. When Hertz places two Riess spirals one above the other and discharges a Leyden jar into one of the Riess spirals, there is a spark that jumps through the air to the other coil! "I had been surprised to find that it was not necessary to discharge large batteries through one of these spirals in order to obtain sparks in the other; that small Leyden jars amply sufficed for this purpose, and that even the discharge of small induction-coil would do, provided it had to spring across a spark-gap." So, in 1887 he designs an experiment in which he uses a transmitter and a receiver. The transmitter is composed of a Leyden jar, an induction coil and a spark gap. The receiver is simply a looped wire with small knobs at the two extremities separated by a gap (see Figure 3.7). When he discharges the jar, he notices that there is a spark that forms between the two knobs of the receiver! This

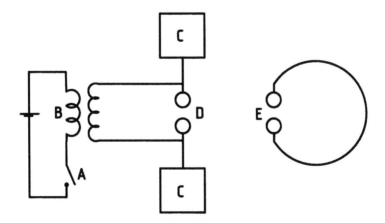

FIGURE 3.7   Hertz's experiment (A: switch; B: transformer; C: metal plates; D: spark gap; E: receiver).

means that the spark produced at the transmitter has propagated through air to reach the receiver! According to him, this experiment shows that Maxwell's equations are valid! "This is just an experiment that proves Maestro Maxwell was right – we just have these mysterious electromagnetic waves that we cannot see with the naked eye. But they are there". Hertz also manages to measure the velocity of these electromagnetic radiations and finds it to be the same as the light's velocity! He shows that the same laws as for light, including reflection and refraction, apply to these electromagnetic waves! His experiment inspires Marconi and Braun, who soon build on it to invent the wireless telegraph, the radio and the television! Both Marconi and Braun receive the Nobel Prize in physics in 1909 for their "contributions to the development of wireless telegraphy". Also, let us add that until around 1910, these electromagnetic waves were called "Hertzian waves". The term "radio waves" becomes commonly accepted later on. In recognition of his work, the unit of frequency for a wave – one cycle per second – is named the "Hertz" (Hz).

Another electromagnetic radiation is discovered in 1895. Röntgen (1845–1923) discovers the "X-rays" (X stands for an unknown in mathematics) or Röntgen rays. While working with a cathode-ray tube in his laboratory, Röntgen observes a fluorescent glow on a paper plate covered with barium platinocyanide, $Ba[Pt(CN)]_4$, even if the paper is 2 meters away from the tube. He then places several objects of different thicknesses between the tube and the paper plate and finds that the solid in the paper plate still glows! Then, he decides to do the following experiment. He uses a photographic plate and places his wife's hand between the tube and the plate; they see her bones and her wedding band! (See Figure 3.8.) It thus means that X-rays can penetrate human flesh, but not dense materials like bones. Röntgen's discovery rapidly has a tremendous impact on medicine since it allows doctors to see inside the human body without performing any surgery! He receives the first Nobel Prize in Physics in 1901. In recognition of his work, element 111 in the periodic table is named roentgenium (Rg).

On February 26 and 27, 1896, to better define these famous X-rays, Becquerel exposes a fluorescent mineral, potassium uranyl sulfate, $K_2UO_2(SO_4)_2$, to sunlight and then places it on photographic plates wrapped in black paper. According to him, uranium will absorb the sun's energy and will then emit it as X-rays. However, the sky is cloudy in Paris during these days. Yet, he decides to develop his photographic plates. To his amazement, he discovers clear pictures as if uranium did not need any sunlight to emit radiation! This opens the door to a new field, radioactivity, highlighting the fact that an element can also emit electromagnetic radiation on its own. Indeed, the Curies start to extract uranium from ore and, to their great surprise, find that the ore is actually more active than pure uranium, meaning that the ore also contains other radioactive elements! This leads them to the discoveries of the elements polonium (Po) (Marie Curie was born in Poland) and radium (Ra) (Latin: *radius* = ray).

However, a challenge remains: the famous blackbody radiation introduced by Kirchhoff. A blackbody is a theoretical model that absorbs all radiations it is subjected to. Hence the term blackbody, because the object would appear perfectly black since it absorbs any incoming light. The energy absorbed results in vibrations of the molecules of the blackbody. The blackbody then releases this energy by emitting thermal radiation known as "blackbody radiation". In 1895, Wien and Lummer realize the following experiment. They punch a hole in a completely closed oven and

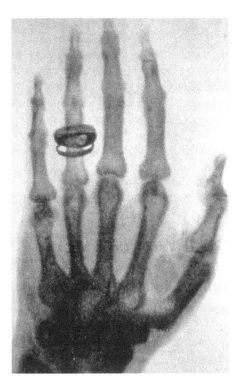

FIGURE 3.8   Röntgen X-rays photography of his wife's hand.

measure, for different temperatures, the characteristics of the radiation that comes out by this hole, including the frequency and the amount of energy released. Wien then formulates a law that correlates the energy with the frequency, for which he receives the Nobel Prize in 1911. He uses classical thermodynamics and finds that the peak frequency is proportional to the temperature. In other words, when the temperature is high, the peak frequency is high. For example, at T = 7,000 K, the peak frequency is 7.25×10^{14} Hz (725 THz) and corresponds to a wavelength of 414 nm (violet). These results match the experimental data. However, Wien's law fails for low frequencies.

Around 1900, Raleigh and Jeans formulate the well-known Rayleigh–Jeans law. It suffers from the converse issue. Although it works better than Wien's formula at low frequencies, it does not explain the high-frequency data. In fact, the Rayleigh–Jeans law even diverges at high frequency, causing the famous ultraviolet catastrophe! In fact, according to this law, there is never a peak frequency, meaning that, regardless of its temperature, an object always radiates more energy for the high frequencies (ultraviolet) than for low frequencies (visible). Yet, we see that stars do have a color and there is indeed a peak frequency for each temperature!

However, nothing seems to explain the entire spectrum of a blackbody. Planck (1858–1947) is the first to solve this puzzle by realizing that the energy of light cannot be understood through classical mechanics. Indeed, he assumes that the amount of energy (E) that can be absorbed, or emitted, by a blackbody can only be of the form $E = nh\nu$ where $n = 0, 1, 2...$ meaning that E can only take the following values: $0, h\nu, 2h\nu, 3h\nu...$ Using these values for E, Planck converts the continuous integrals used by Raleigh and Jeans to discrete sums over an infinite number of terms. Making that simple change gives Planck a new formula for the spectrum of the blackbody radiation. And this equation gets it right! It exactly describes the blackbody spectrum, both at low and high frequencies! Planck's idea is revolutionary! Planck's equation is now known as Planck's quantization rule, and h is named Planck's constant ($h = 6.626 \times 10^{-34}$ Js). In short, Planck assumes that electromagnetic radiation can only be emitted or absorbed in discrete packets, called quanta of energy ($h\nu$)! This is the birth of quantum physics! Let us add that the concept of blackbody has been proved to exist. Indeed, the cosmic microwave background (CMB) radiation, emitted 380,000 years after the Big Bang, shows a spectrum that matches perfectly the blackbody radiation at a temperature of 2.73K. This is discovered by Mather and Smoot, who receive the Nobel Prize in Physics in 2006.

DISCOVERY OF THE ELECTRON AND FIRST ATOMIC MODELS

During the 19th century, a new type of entertainment makes its debut in theaters: magic. One of the most prominent illusionists at the time is Robert-Houdin (1805–1871). He is trained as a watchmaker, but becomes keenly interested in magic after reading the famous "Scientific Amusements" written in 1790 by Tissandier. "Young people of both sexes, and persons of all ages who have leisure and a taste for that which is ingenious as well as instructive and amusing, may be commended to this remarkably interesting collection of experiments, nearly all of which can be readily performed by an unskilled person who will carefully follow out the directions given. It is surprising how near we are to the most fundamental principles of science when we perform some of the simplest operations." Indeed, from the Preface, one can see that magic is closely related to science. This book reviews principles of physics, such as gravitation, hydrostatics, heat conduction, optics, electricity and magnetism, and shows how they can be applied to create illusions. Robert-Houdin is the first magician to use electricity in his tricks, and he pushes the boundaries of magic further by creating his own illusions, including the "floating boy" (levitation trick using the mysterious "ether") and the "marvelous orange tree". By the way, the famous illusionist Harry Houdini takes his name as an homage to Houdin, and he also owns a copy of "Scientific Amusements"! Let us add that magicians also take a lot of interest in the invention of the cathode-ray tube and in the fluorescent glows it creates. Indeed, in the 1850s, Pluecker and his student Hittorf run an electric current in a tube which contains a rarefied gas. They observe a fluorescent glow! To better understand this phenomenon, they move a magnet close to the tube and see that the glow shifts. According to them, this glow is, in fact, a charged "bundle of particles". This finding is confirmed in 1876 by Goldstein, who calls this glow "*Kathodenstrahlen*" or "cathode rays". Meanwhile, in England, Crookes develops his own cathode-ray tubes and describes them as "molecules of the gaseous residue in highly exhausted vessels that are able to dart across the tube with comparatively few collisions and which radiate from the pole with enormous velocity" (see Figure 3.9). He gives them the name of

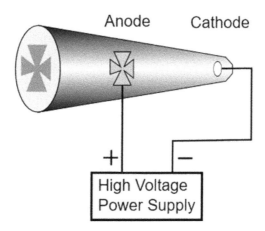

Anode   Cathode

FIGURE 3.9   Crookes' cathode-ray tube.

"radiant matter", believing that he has found a fourth state of matter! This becomes an exciting research topic studied all over Europe.

In Cambridge (UK), another famous Thomson, J. J. Thomson (1856–1940) has a revolutionary idea. According to him, cathode rays are not just material particles but they are, in fact, building blocks of atoms and the elusive basic unit of all matter in the universe! To demonstrate this, he performs three experiments. First, he builds on an experiment carried out by Perrin (1870–1942). In 1895, Perrin adds a metal cylinder at the end of the cathode ray tube to collect the mysterious cathode rays. Perrin connects the metal cylinder to an electroscope, which is a device that measures electrical charges. He finds that cathode rays deposit negative charges! Thomson wants to see if, by bending the cathode rays with a magnet, he can separate the charges from the rays. Thus, he modifies Perrin's experiment and adds a magnet to bend the cathode rays before they enter the metal cylinder. Thomson finds that, if the rays enter the cylinder, the electroscope registers a large amount of negative charge. On the other hand, if the rays are deflected and do not enter the cylinder, no charge is recorded by the electroscope. According to Thomson, negative charges and cathode rays are somehow stuck together: You cannot separate the charges from the rays!

Still interested in isolating the charges, he thinks of using an electric field. So far, several distinguished scientists, including Hertz, have not managed to achieve this. Thomson thus designs a second experiment. The idea is to extract most of the gas from the tube. By doing so, he finds that the cathode rays do bend in an electric field after all! Previous failed attempts were due to the presence of too many gas molecules in the tube that acted as conductors. Thomson concludes from these two experiments, "I can see no escape from the conclusion that [cathode rays] are charges of negative electricity carried by particles of matter". However, he continues, "What are these particles? Are they atoms, or molecules, or matter in a still finer state of subdivision?" The aim of Thomson's third experiment is to determine the properties of the particles. He performs a series of experiments using different gases and tubes. He stores data such as how much the rays are deflected by a magnetic field and how fast the rays travel. From the data, he calculates the ratio e/m (electric charge of the particle over its mass)! Thomson states: "We have in the cathode rays matter in a new state, a state in which the subdivision of matter is carried very much further than in the ordinary gaseous state: a state in which all matter – that is, matter derived from different sources such as hydrogen, oxygen, [...] – is of one and the same kind; this matter being the substance from which all the chemical elements are built up".

At the time, only the quantity e/m is known and the exact charge (e) of this corpuscle still remains an enigma. In 1909, Millikan (1868–1953) designs the famous "falling-drop method" to determine the charge of this corpuscle (see Figure 3.10). The idea is to suspend charged oil droplets between two metal electrodes by imposing an upward electric force that balances the downward gravitational force. Knowing the value of the electric field and the mass of the droplet, he determines the charge to be 1.5924×10^{-19} Coulomb (within 1% of the present-day constant of $1.602176487 \times 10^{-19}$ C)! Using the measured charge of an electron, one can now calculate the mass of this corpuscle through the e/m ratio found in Thomson's cathode ray experiment and obtain $m = 9.1 \times 10^{-28}$ g.

Let us add that Thomson is awarded the Nobel Prize in 1906 "in recognition of the great merits of his theoretical and experimental investigations on the conduction of electricity by gases" and Millikan in 1923 "for his work on the elementary charge of electricity". Furthermore, Thomson becomes a strong advocate of the "plum-pudding" or "raisin bread" model to represent the atom. The idea is that plenty of tiny, negatively charged, corpuscles (plums) are scattered inside a positively charged fluid (pudding).

Let us add that this corpuscle becomes known later as the electron! The discovery of the electron prompts a series of new technologies that change the world! For instance, the television sets and computer monitors of the 20th century are all descendants of the cathode-ray tube that Thomson uses in his 1897 experiment. Moreover, this leads to the invention of electron microscopy. Unlike a conventional

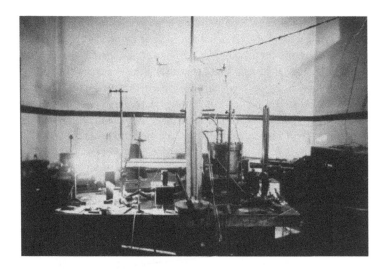

FIGURE 3.10 Millikan's falling drop method.

microscope, the light source is replaced by a beam of accelerated electrons. Indeed, since the wavelength of an electron is many orders of magnitude shorter than the light used in a regular microscope, electron microscopes (EM) allow for a much higher resolution, revealing the structure of very tiny objects. In 1939, Ruska, Kausche and Pfankuch are the first to visualize a virus with an EM! In 1940, the transmission electron microscope (TEM) is invented. In this case, the electron beam travels through the sample and allows an image to be obtained of objects that are up to 1 μm (1×10^{-6} m) in thickness! Then, the scanning electron microscope (SEM) is invented and provides an image of the surface of an object, giving access to the depth or 3D map of the system studied! In 1986, Ruska (1906–1988) shares the Nobel Prize with Binning and Rohr, developers of the scanning tunneling electron microscope (a combination of SEM and TEM). This new device makes it possible to follow very precisely what happens structurally on the surface, one atom at a time! And the story does not end here. In 2017, Dubochet, Frank and Henderson are awarded the Nobel Prize in Chemistry "for developing cryo-electron microscopy for the high-resolution structure determination of biomolecules in solution". This invention enables the visualization of 3D images of any biomolecule, including the surface of the Zika virus!

RADIOACTIVITY AND RUTHERFORD'S NUCLEAR MODEL

In 1895, Rutherford (1871–1937) starts working at the Cavendish Laboratory with Thomson. His project consists of using a cathode-ray tube to study the effect of the recently discovered X-rays on gases. A year later, he publishes an article "On the Passage of Electricity through Gases exposed to Roentgen Rays" in which he shows that a gas subjected to X-rays becomes an electricity conductor! This is an impressive result as a gas is normally an insulator (does not conduct electricity). This also means that charges are produced! Specifically, he states that "The term ion was given to them from analogy with electrolytic conduction, but in using the term it is not assumed that the ion is necessarily of atomic dimension; it may be a multiple or submultiple of the atom". This explains the conductivity of the gas! He adds "The positive and negative ions will be partially separated by the electric field, and an excess of ions of one sign may be blown away, so that a charged gas will be obtained". At the same time, he becomes aware of Becquerel's investigations on uranium. "The results of Becquerel showed that Roentgen and uranium radiations were very similar in their power of penetrating solid bodies and producing conduction in a gas exposed to them". In 1899, Rutherford carries out a series of experiments to study the absorption of rays emitted by uranium. Between a compound of uranium and an electroscope, he places a single aluminum foil, then two, and so on, to study the effect of thickness on uranium rays. At each stage, he measures the intensity of the radiation. He finds two types of radiation that differ in their ability to penetrate matter. The first type of radiation is very easily absorbed by a few layers of foil, while the second seems to go through up to the 100th aluminum foil! He calls them α and β rays "for convenience". He later publishes his main results in "Radio-activity" (1904) and "Radioactive Transformations" (1906), in which he examines the "amount of ionization, [...] due to the α, β, and γ rays is of the relative order 10,000, 100, and 1". He also manages to calculate the "thickness of aluminum which cuts off half the radiation" and finds the numbers of 0.0005 cm for α, 0.05 cm for β and 8 cm for γ. Let us add that Rutherford coins the term γ rays in 1903 for rays that have more penetrating power than α and β rays.

At the same time, Ramsay and Soddy analyze the gases produced by a solution of radium bromide in water. Using spectroscopy analysis, they find that "the D_3 line of helium appeared"! This is a very surprising result, since the only elements present initially are hydrogen and oxygen from water, as well as radium and bromine, but not helium, bringing back the alchemical concept of transmutation and also prompting new questions for Rutherford: "Has the α particle expelled from all radioactive products the same mass? Does the value of e/m of the α particle vary in its passage through matter? What is the connection between the velocity of the α particle and its range of ionization in air? What is the connection, if any, between the α particle and the helium atom?" To better answer these questions, Rutherford and Geiger decide to develop new experimental techniques for the detection and measurement of α particles. Indeed, since the early experiments of Rutherford on radioactivity, they use a method relying on the "well-known property of producing scintillations in a preparation of phosphorescent zinc sulfide. With the aid of a microscope, it is not very difficult to count the number of scintillations appearing per second on a screen of known area when exposed to a source of α rays." However, this method is subject to human error in the count, so they finally decide to design the famous "Geiger counter". The experimental device is explained in a famous paper entitled "An Electrical Method of Counting the Number of α-Particles from Radio-active Substances" in 1908. They use the "principle of production of fresh ions by collision" introduced by Townsend. The idea is to use a strong electric field to produce ions by collisions with molecules of a neutral gas. He notes that, beyond a threshold for the electric field, the current rises rapidly. According to Townsend, "this effect is due to the production of fresh ions in the gas by the collision of the negative ions with the gas molecules". Therefore, each α particle produces a cascade of ions, which results in a sudden current and allows the electrometer to detect accurately the passage of an α particle! Geiger concludes that "in our experiments to detect a single α particle, it was arranged that α particles could be fired through a gas at low pressure exposed to an electric field somewhat below the sparkling value. In this way, the small ionization produced by one α particle in passing along the gas could be magnified several thousand times." Using this new way of counting, Rutherford is able to answer the most pressing question in radioactivity: Is an α particle an atom of helium? And the answer is: almost. He finds that an α particle is, in fact, a helium atom with a charge of 2e. In today's terms, one would say a helium atom stripped of its electrons. In radiochemistry, α particle can also be written as $_2^4He^{2+}$ or He^{2+} (to be compared to the helium atom $_2^4He$). In 1908, Rutherford is awarded the Nobel Prize in Chemistry "for his investigations into the disintegration of the elements, and the chemistry of radioactive substances".

In 1909, Rutherford's research group performs the famous gold foil experiment (see Figure 3.11) and discovers the proton! The experimental device is the following. They bombard a beam of α particles on an ultrathin gold foil and then detects the scattered α particles on a ZnS screen. They find that most of the particles go through the foil without being deflected, some of the α particles are deflected with a

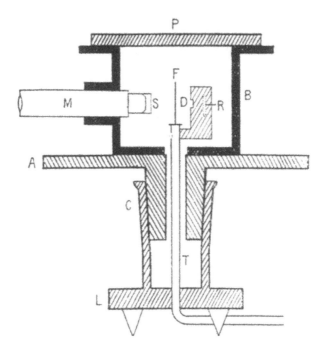

FIGURE 3.11 Gold foil experiment: Geiger–Marsden apparatus.

very small angle and a very few bounce back (1 in 20,000). Based on their observations, Rutherford "supposed that the atom consisted of a positively charged nucleus of small dimensions in which practically all the mass of the atom was concentrated. The nucleus was supposed to be surrounded by a distribution of electrons to make the atom electrically neutral, and extending to distance from the nucleus comparable with the ordinary accepted radius of the atom." In other words, Rutherford suggests a new atomic model in which the atom looks like a very small solar system, with a massive, positively charged center surrounded by electrons! Let us add that, in 1919, Rutherford designs a very interesting experiment to study another radioactive material: radium C (bismuth-214 or $^{214}_{83}Bi$) in different gases. He describes the following finding for nitrogen: "From the results so far obtained, it is difficult to avoid the conclusion that the long-range atoms arising from collision of alpha particles with nitrogen are not nitrogen atoms but probably atoms of hydrogen." He adds that "we must conclude that the nitrogen atom is disintegrated under the intense forces developed in a close collision with a swift alpha particle, and that the hydrogen atom which is liberated formed a constituent part of the nitrogen nucleus." In other words, when an α particle ($^4_2He^{2+}$) collides with a nitrogen atom ($^{14}_7N$), nitrogen is disintegrated! And there is the emission of a "swift hydrogen atom" (now called a proton)! Now, we would translate this phenomenon into the following equation or nuclear reaction: $^{14}_7N + ^4_2\alpha \rightarrow ^{17}_8O + ^1_1p$. As we can see, $^{14}_7N$ is disintegrated into $^{17}_8O$!

Moreover, several experiments indicate that the mass of the nucleus is roughly equal to two times the mass of the protons it contains, meaning that there is an additional mass. Rutherford thus assumes that there are also neutral particles in the

nucleus, with a mass similar to protons. However, there are no experimental confir-
mations yet. Many scientists decide to tackle this problem. In 1932, Chadwick builds
on an experiment developed by Irene and Frederic Joliot-Curie. The main difference
in his approach is that he has better detection equipment! He finds that "the properties of the penetrating radiation emitted from beryllium [...] when bombarded by
the α particles of polonium have been examined. It is concluded that the radiation
consists, not of quanta as hitherto supposed, but of neutrons, particles of mass 1, and
charge 0. Evidence is given to show that the mass of the neutron is probably between
1.005 and 1.008." In radiochemistry, beryllium undergoes the following reaction
when it is bombarded with an α particle: $^9_4Be + ^4_2\alpha \rightarrow ^{12}_6C + ^1_0n$.

These discoveries prompt the design of the first particle accelerator to produce nuclear disintegrations. In 1932, Cockcroft and Walton accelerate protons on
a lithium target. They split the lithium nucleus into two alpha particles! $^1_1p + ^7_3Li$
$\rightarrow ^4_2\alpha + ^4_2\alpha$. Let us also add that, in 1932, Blackett and Occhialini develop the cloud
chamber that they connect to a Geiger counter to detect the passage of a particle.
They discover the existence of the "positive electron", also known as a positron.
Indeed, light particles, when subjected to a high energy, can give rise to pairs of
electrons and positrons: $\gamma \rightarrow e^- + e^+$, a first sign of the existence of antimatter!

Rutherford is seen as the father of nuclear physics and is conferred the title of
Baron Rutherford of Nelson in 1931. He is buried in London at Westminster Abbey,
alongside Newton and Kelvin. If you travel to New Zealand, you will see his picture on the $100 New Zealand bills. Finally, the chemical element 104 is named
Rutherfordium (Rf) after him.

But the story does not end here. In 1968, deep inelastic scattering experiments at
the Stanford Linear Accelerator Center prove the existence of quarks, predicted in
1964 by Gell-Mann and Zweig. Protons and neutrons are, in fact, composed of three
quarks each! Quarks are eventually found to come in six types or flavors: up, down,
charm, strange, top and bottom.

WAVE–PARTICLE DUALITY: A REALITY

On the theoretical side, maybe one of the most important achievements is due to
Planck (1858–1947). Planck studies for several years the blackbody problem enunci-
ated by Kirchhoff. Planck explains in his Nobel lecture: "Since Gustav Kirchhoff
has shown that the state of the heat radiation which takes place in a cavity bounded
by any emitting and absorbing substances of uniform temperature is entirely independent upon the nature of the substances, a universal function was demonstrated
which was dependent only upon temperature and wavelength, but in no way upon the
properties of any substance." He finds something that will revolutionize the entire
physics world! His goal is to identify this function by gathering many experimental
results and finding a universal relation using mathematics. Planck develops a model
for a blackbody by using several Hertz linear oscillators to mimic the blackbody
radiation within its cavity. Indeed, according to Maxwell's theory, the oscillators
emit electromagnetic waves. Thanks to this model, he obtains the first laws of emission and absorption of a linear resonator and checks them with experimental measurements on vapors. It works! There is a "general connection between the energy

of a resonator of specific natural period of vibration and the energy radiation of the corresponding spectral region" also meaning that "the energy of the resonator [...] could be replaced by a simple system of one degree of freedom". This greatly simplifies the problem! He then decides to tackle the problem from the standpoint of thermodynamics. He attempts to find a relation between the energy of the resonator and its entropy or, more precisely, with the second derivative of entropy which measures the irreversibility of the energy exchange. He uses Wien's experimental results to go further: A "law brought out the dependence of the radiation intensity on the temperature, representing it by an exponential function. If one calculates the connection between the entropy and the energy of a resonator, [...] the remarkable result is obtained that the reciprocal value of the above-mentioned differential coefficient, which I will call R, is proportional to the energy." Other experiments allow him to define R more precisely. He notices that, for small energies, the function R is proportional to the energy. On the other hand, for large energies, R varies according to the square of the energy. He thus writes the function R as the sum of these two terms. However, Planck is not completely satisfied with this radiation formula; it still has a limited value, in the sense that this is just an interpolation formula. "For this reason, I busied myself, from then on, that is, from the day of its establishment, with the task of elucidating a true physical character for the formula, and this problem led me automatically to a consideration of the connection between entropy and probability, that is, Boltzmann's trend of ideas; until after some weeks of the most strenuous work of my life, light came into the darkness, and a new undreamed-of perspective opened up before me." Using statistical thermodynamics, Planck finds that the probability of a specified energy distribution in a system of resonators leads to the same entropy expression as when using the radiation law, meaning that the statistical way can be a solution! Planck's radiation law gives the energy distribution of electromagnetic radiation in thermodynamic equilibrium with matter at a given temperature. The energy density ρ (energy of the radiation per unit volume) at a given frequency μ and temperature T is then given by $\rho_{\upsilon,T} = 8\pi h\upsilon^3/c^3 \times 1/(\exp(h\upsilon/k_B T) - 1)$. As we can see, this equation contains two constants k_B. The first constant, nowadays known as the Boltzmann constant k_B, is connected to the definition of temperature. Let us recall that temperature is also related to the kinetic energy. For instance, for one mole of an ideal gas, E_k is given by 3/2RT which is also equal to $3/2N_a k_B T$, meaning that k_B can be calculated through $k_B = R/Na$! Incidentally, Boltzmann was never interested in carrying out an exact measurement of the constant k_B. The second constant is h. According to Planck himself, "The explanation of the second universal constant of the radiation law was not so easy. Because it represents the product energy and time [...], I described it as the elementary quantum of action". The product $h\upsilon$, where υ is the frequency of vibration of the radiation, is actually the smallest amount of heat which can be radiated at the vibration frequency υ, though the famous E = hυ. However, it also raises a few questions. "Either the quantum of action was a fictional quantity, [...] or the derivation of the radiation law was based on a sound physical conception. In this case the quantum of action must play a fundamental role in physics, and here was something entirely new, never before heard of, which seemed called upon to basically revise all our physical thinking, built as this was, since the establishment of the infinitesimal calculus by Leibniz and Newton, upon the acceptance of the continuity of all causative connections." He also adds "to be sure, the introduction of the quantum of action has not yet produced a genuine quantum theory". Planck's formula leads him to a value for Avogadro's number of molecules in a mole (the gram molecular weight) of a gas and to an estimate of the fundamental unit of electrical charge. These give Planck great confidence that his "fictitious" formula must be correct! Planck receives the Nobel Prize in Physics in 1918 "in recognition of the services he rendered to the advancement of Physics by his discovery of energy quanta".

Another proof of the validity of this theory comes from the Rydberg formula, which can be used to calculate the spectral emission lines of atomic hydrogen. Indeed, using Bohr's atomic model, several different series of emission lines can be defined. According to Bohr, an atom is composed of electrons and of a nucleus. Electrons revolve around the nucleus on specific orbits. The distance between the nucleus and an electron on a specific orbit can be found using the following formula: $n\lambda = 2\pi r$. For example, for the closest orbit $n = 1$, the distance between the nucleus and the electron is 0.0529 nm, also known as the Bohr radius. Bohr proposes to calculate the energies of these specific orbits for the hydrogen atom and other hydrogen-like atoms and ions. Indeed, electrons can only gain and lose energy by jumping from one orbit to another, absorbing or emitting an electromagnetic radiation of frequency ν determined by the Planck relation: $\Delta E = E_2 - E_1 = h\nu$. Nowadays, we use the Rydberg formula for hydrogen-like atoms, which is a generalization of the previous formula: $1/\lambda = R_H(1/n_1^2 - 1/n_2^2)$ with n_1 the principal quantum number for the upper energy level and n_2 the corresponding quantum number for the lower energy level of the atomic electron transition. For $n_1 = 1$ and $n_2 = 2 \rightarrow \infty$, the series is called the Lyman series (see Figure 3.12). For $n_1 = 2$ and $n_2 = 3 \rightarrow \infty$, it is called the Balmer series. $n_1 = 3$ and $n_2 = 4 \rightarrow \infty$ gives the Paschen series. Bohr receives the Nobel Prize 1922 "for his services in the investigation of the structure of atoms and of the radiation emanating from them".

Planck's theory draws more and more attention. In 1927, during the 5th Solvay conference in Brussels, the foremost physicists in the world focus on "Electrons and Photons". Around the chair Hendrik Lorentz, we find Piccard, Henriot, Ehrenfest, Herzen, de Donder, Schrödinger, Verschaffelt, Pauli, Heisenberg, Fowler, Brillouin,

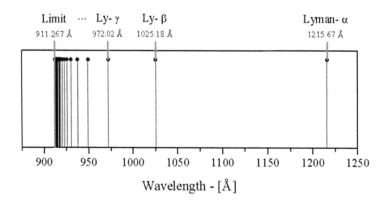

FIGURE 3.12 Lyman series.

Debye, Knudsen, Bragg, Kramers, Dirac, Compton, de Broglie, Born, Bohr, Langmuir, Planck, M. Curie, Einstein, Langevin, Guye, Wilson and Richardson. Let us note that 17 of the 29 attendees are or will become Nobel Prize laureates!

De Broglie thinks of a new concept: a wave nature for the electron. According to him, "for both matter and radiations, light in particular, it is necessary to introduce the corpuscle concept and the wave concept at the same time". His idea is to associate the motion of a corpuscle to the propagation of the corresponding wave. "A wave must be associated with each corpuscle and only the study of the wave's propagation will yield information to us on the successive positions of the corpuscle in space". Mathematically, this translates into the following relation: $\lambda = h/p$ that links the group velocity of the waves (through λ) to the velocity of the corpuscle (through p). This notion of group velocity means that "it is not constantly possible to assign to the corpuscle a well-defined position in the wave". More specifically, he adds that: "The electron can no longer be conceived as a single, small granule of electricity; it must be associated with a wave and this wave is no myth; its wavelength can be measured and its interferences predicted. It has thus been possible to predict a whole group of phenomena without their actually having been discovered. And it is on this concept of duality of waves and corpuscles in Nature, expressed in a more or less abstract form, that the whole recent development of theoretical physics has been founded and that all future development of this science will apparently have to be founded." De Broglie receives the Nobel Prize in 1929 "for his discovery of the wave nature of electrons". Meanwhile, Einstein explains the photoelectric effect using "light quanta" or discrete units of light energy, later called photons.

PHOTOELECTRIC EFFECT AND DISCOVERY OF THE PHOTON

In 1921, Einstein is awarded the Nobel Prize in Physics "for his services to Theoretical Physics, and especially for his discovery of the law of the photoelectric effect". This discovery is very important because it shows that, under specific circumstances, light does not behave as a wave but as a particle! It also provides a first example of the concept of wave–particle duality, which will become the focus of a new scientific field known as quantum mechanics. Einstein states that light is composed of "light quanta" (packets) of fixed energies. Let us add that later on, in 1926, these quanta will be named photons by Lewis (1875–1946): "I therefore take the liberty of proposing for his hypothetical new atom, which is not light but plays an essential part in every process of radiation, the name of photon". The idea is that different "packets" have different energies, $E_i = h\nu_i$, depending on their frequencies (ν_i). In this equation, E_i is the energy of a particle i representing a quantum of light (photon) and ν_i the frequency of an electromagnetic wave. This relation is fundamental since it shows the wave–particle duality of light! Einstein uses Planck's concept of "quanta" of energy but, unlike Planck, does not think that this is a mathematical trick!

When light bounces on a metal, photons interact with the electrons in the metal. This collision between a photon and an electron can create a "photoelectric effect" which, in turn, produces electric sparks! Indeed, a photon brings an energy $E = h\nu$ that serves to extract an electron from the metal and gives a kinetic energy to this electron. According to Einstein, since $h\nu = E_{extraction} + E_{kinetic}$, there is a photoelectric effect if $h\nu > E_{extraction}$! In other words, when $h\nu < E_{extraction}$, or $\nu < (E_{extraction}/h)$, or equivalently $\nu < \nu_{threshold}$, there is no photoelectric effect! If the theory seems to work on paper, experimental evidence has yet to come.

The answer to this question is given by Millikan in his 1914 experiment. Let us add that he also receives the 1923 Nobel Prize in Physics "for his work [...] on the photoelectric effect". He starts from Einstein's proposition: $h\nu = E_{extraction} + E_{kinetic}$. He decides to replace $E_{kinetic}$ by $\frac{1}{2}mv^2$ and $E_{extraction}$ by P leading to $\frac{1}{2}mv^2 = h\nu - P$, where h is Planck's constant, ν the frequency of the incident light and P is, in Millikan's words, "the work necessary to get the electron out of the metal". Millikan's idea is to show that this equation is invalid. In 1923, he states that "This work resulted, contrary to my own expectation, in the first direct experimental proof [...] of the Einstein equation, and the first direct photo-electric determination of Planck's h". Indeed, Millikan starts with the idea that the more intense the beam of the light is, the more electrons are ejected from the metal, no matter what the frequency is. However, as Einstein predicted, he finds that light with frequency $\nu < \nu_{threshold}$ does not eject electrons from the metal surface! Moreover, using a stopping potential to counteract the kinetic energy of the electron ejected from the metal surface, he discovers that he manages to stop the current in his experiment! The value of this stopping potential (V_0) gives a measure of the maximum kinetic energy of the electrons resulting from the photoelectric effect. Indeed, we can replace $E_{kinetic}$ by eV_0 in the equation $h\nu = E_{extraction} + E_{kinetic}$, and we obtain $h\nu = h\nu_{threshold} + eV_0$. Now, if we divide the equation by h, we obtain $\nu = \nu_{threshold} + (e/h)V_0$. If one plots a graph representing different frequencies for the incident light with respect to the applied V_0, one obtains lines with slopes of (e/h) (see Figure 3.13). And since e and h are constants, the different curves must have the same slope! By the same token, it demonstrates that h is constant and thus that the formula for quantum energy $E = h\nu$ is actually very true! Millikan finds that everything discussed above holds. The light is thus both a particle and a wave!

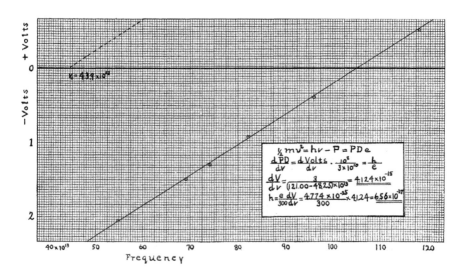

FIGURE 3.13 Millikan's 1914 experiment.

Nowadays, solar cell technology relies on Einstein's photoelectric effect. A solar cell, or photovoltaic cell, is composed of two layers of a semiconductor material like silicon. The top layer is doped in electrons (rich in electrons) while the bottom layer is poor in electrons. When light is beamed on the top layer, it frees and pushes an electron that moves towards the bottom layer, creating a current. This current is then used, for example, to power engines for electric cars. Let us add that, in 1921, when Einstein receives the Nobel Prize for the photoelectric effect, his Nobel lecture is focused on another topic: relativity!

Special relativity is introduced in 1905 in "On the Electrodynamics of Moving Bodies". It relies on two principles. "1. The laws by which the states of physical systems undergo change are not affected, whether these changes of state be referred to the one or the other of two systems of co-ordinates in uniform translatory motion". "2. Any ray of light moves in the 'stationary' system of co-ordinates with the determined velocity c, whether the ray be emitted by a stationary or by a moving body". Special relativity refers to the case where motion is uniform. More specifically, it focuses on the special case of objects traveling along a straight line and at a constant velocity. In a nutshell, special relativity applies to all systems for which gravitation does not play a major role, leading to the concept of "flat spacetime". Moreover, the energy of an object (E) can be calculated as a function of its mass (m) and momentum (p) through $E^2 = p^2 c^2 + m^2 c^4$. For an object at rest (p=0) we recover the famous equation: $E = mc^2$! By doing so, Einstein shows that mass is also a type of energy! Indeed, his theory implies that an object, even with a very small mass, can give rise to a tremendous amount of energy (as it is multiplied by c^2 with $c = 3 \times 10^8$ m/s)! This is at the heart of nuclear energy. Indeed, nuclear power plants use a process known as nuclear fission to transform a mass change into energy production. In particular, in a nuclear reactor, neutrons are bombarded on uranium atoms $^{235}_{92}U$ to form $^{236}_{92}U$. $^{236}_{92}U$ split into $^{139}_{56}Ba$ and $^{94}_{36}Kr$, releasing three neutrons ($^1_0 n$) and resulting in a final product that is lighter than the initial product or, in other words, in a "mass defect". This change in mass Δm results in an energy production of $E = \Delta mc^2$ of 210 MeV (210,000,000,000 eV) per atom! To give a better idea of the gigantic amount of energy involved, 1 kg of uranium can yield 9×10^{16} J, the equivalent of burning 3 million tons of coal!

However, special relativity does not apply any longer to objects that are accelerating or changing trajectories; this is the reason why, in 1915, Einstein develops a general theory of relativity. He determines that objects with a large mass, such as planets, cause a distortion in spacetime! This is manifested in the phenomenon of gravity. He also proposes the concept of time dilation which relies on both special and general relativity. There is a famous thought experiment known as the "twin paradox". In this experiment, one of the twins stays on the Earth while the other embarks a spaceship to travel around the Earth. Because of the time-dilation effect, the time goes slower on the spaceship! Indeed, when the spaceship lands back on Earth, the twin who was on the spaceship has aged less than the twin who stayed on the Earth! This creates great excitement, and prompts the foremost intellectuals of the beginning of the 20th century to consider the following question: What is time? A famous meeting between Einstein and Bergson (1927 Nobel Prize in Literature) takes place in Paris in 1922. Bergson is convinced that Einstein's theory is not consistent with our intuition of time, which is an intuition of duration. Bergson's theory of duration is exposed in

"Time and Free Will: An Essay on the Immediate Data of Consciousness" in which he studies the question: what is time? "If I glance over a road marked on the map and follow it up to a certain point, there is nothing to prevent my turning back and trying to find out whether it branches off anywhere. But time is not a line along which one can pass again." He imagines the dialogue between defenders and opponents of free will: "The former reason thus: 'The path is not yet traced out, therefore it may take any direction whatever.' To which the answer is: 'You forget that it is not possible to speak of a path till the action is performed: but then it will have been traced out.' The latter say 'The path has been traced out in such and such a way: therefore its possible direction was not any direction whatever, but only this one direction.' To which the answer is: 'Before the path was traced out there was no direction, either possible or impossible, for the very simple reason that there could not yet be any question of a path'." After a 30-minute presentation by Bergson, Einstein replies: "There is no 'philosopher's time' which is both psychological and physical; there exists only a psychological time which is different from the time of the physicist"!

Finally, the reality of time dilatation was demonstrated by using atomic clocks on the NASA Space Shuttle Challenger in 1985. Indeed, these clocks were running slightly slower than reference clocks on Earth! Another illustration of time dilation is the phenomenon of gravitational lensing. Indeed, a light around a massive object, such as a black hole, is bent, transforming it into a lens for the objects that are behind it. Nowadays, astronomers use this technique to study stars and galaxies behind massive objects, but also dark matter, supermassive black holes and even to determine if the universe will forever expand or eventually collapse!

FIRST QUANTUM-BASED THEORY

The beginning of the 20th century is marked by tremendous new scientific developments and specifically by the birth of quantum mechanics. Let us quickly recap the key achievements made in just a few decades. It starts with Planck and Einstein, who introduce and apply the concept of quanta to characterize the energy of light. Indeed, matter can only emit or absorb light through specific amounts of energy, whose values are multiples of a quantum. Moreover, their work highlights that, for any color of the light spectrum, the ratio of the energy to the frequency is always equal to a universal constant: Planck's constant, h! It also connects the energy of a particle to the frequency of a wave, emphasizing the particle–wave duality! De Broglie illustrates further this concept by associating the velocity of a material point to the group velocity of a wave. In other words, a material point can be represented as a wave crest, having a velocity equal to the group velocity, also known as a "wave packet" (see Figure 3.14). From there, quantum mechanics develops into a very complex and profound theory that forever changes the scientific world. In particular, it provides access to a world that is beyond our perception and senses, only accessible through mathematical equations!

Heisenberg (1901–1976) introduces a new formalism, known as matrix-mechanics, that can handle systems of electrons, atoms and molecules. Applying this theory to the hydrogen molecule, he finds that there exist, in fact, two different forms of hydrogen molecules, one called ortho-hydrogen and the other para-hydrogen. At

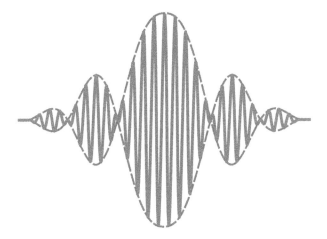

FIGURE 3.14 Wave packets.

room temperature, there are three times more ortho- than para-hydrogen molecules. Pure para-hydrogen can only be obtained at 0K. One of his major findings is that quantum mechanics cannot determine exactly, at any given time, both the position and velocity of a particle, meaning that there is indefinite causality. Since we can only discern average values, due to the imperfection of our apparatus but also of our senses, we can only probe nature in terms of probability. Indeed, Heisenberg states: "Closer examination of the formalism shows that between the accuracy with which the location of a particle can be ascertained and the accuracy with which its momentum can simultaneously be known, there is a relation according to which the product of the probable errors in the measurement of the location and momentum is invariably at least as large as Planck's constant divided by 4π. In a very general form, therefore we should have $\Delta p \Delta q \geq h/(4\pi)$." He finally adds, "Classical physics represents that striving to learn about Nature in which essentially we seek to draw conclusions about objective processes from observations and so ignore the consideration of the influences which every observation has on the object to be observed; classical physics, therefore, has its limits at the point from which the influence of the observation on the event can no longer be ignored. Conversely, quantum mechanics makes possible the treatment of atomic processes by partially foregoing their space-time description and objectification." Heisenberg receives the Nobel Prize in 1932 for "the creation of quantum mechanics, the application of which has, inter alia, lead to the discovery of the allotropic forms of hydrogen".

Schrödinger (1887–1961) formalizes the quantum mechanical wave equation. He starts from the wave theory of light and focuses on the meaning of light rays. According to him: "They are not the physical paths of some particles of light, but are a mathematical device, the so-called orthogonal trajectories of wave surfaces, imaginary guide lines as it were, which point in the direction normal to the wave surface in which the latter advances". In particular, he notes that, to comply with the Fermat principle (shortest light time), one needs to take into account these guide lines and not the wave surfaces, phenomena that he calls a "mathematical curiosity". His idea is the following:

Use the analogy between the Fermat principle and the Hamilton principle. In other words, he is interested in finding how an atom (or a subatomic particle) will evolve in the future. Specifically, Schrödinger aims at developing a wave equation for electrons akin to the one used for light. In 1926, he comes up with an equation that actually governs the evolution of the mysterious "matter waves" over time. Its solution would be a wave function ψ (x,y,z,t). For a single particle moving around in 3D, it can be written as

$$(ih/2\pi)\, \partial\psi/\partial t = (-h^2/(8\pi^2 m))\,(\partial^2\psi/\partial x^2 + \partial^2\psi/\partial y^2 + \partial^2\psi/\partial z^2) + V\psi$$

In this equation, $i=\sqrt{-1}$, h is Planck's constant, m the mass of the particle and V the potential energy of the particle (describing the interactions between the particle and its surroundings). Using bra–ket notation, Schrödinger's equation reads $\hat{H}|\psi(t)\rangle = i\hbar\partial/\partial t\, |\psi(t)\rangle$ and its solution is the wave function ψ(x,y,z,t)! At equilibrium or, in other words, under stationary conditions, Schrödinger's equation becomes $\hat{H}|\psi\rangle = E|\psi\rangle$ in which ψ only depends on x,y and z. For a particle trapped in a 1D box of length L, the solution to the equation (wave function) is $\psi(x) = A \sin (kx)$ in which $k=n\pi/L$ and $A = \sqrt{(2/L)}$ (see Figure 3.15). Let us add that, in 1926, Born gives a probabilistic interpretation of ψ. He finds that $|\psi(x,y,z,t)|^2$ is the probability density for finding the particle at position (x,y,z) at time t. Differently put, the probability for the particle to be located in a domain D at time t is given by the integral $\int_D |\psi(x,y,z,t)|^2\, dxdydz$.

Let us add the famous story about Schrödinger's cat. In 1935, Schrödinger proposes a thought experiment in which a hypothetical cat in a box may be "living and dead [...] in equal parts", a state known as a quantum superposition. However, to find if the cat is alive or dead, the observer has to open the box and, therefore, to interfere

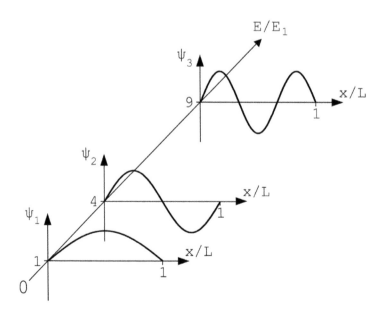

FIGURE 3.15 Wave function for an electron in 1D box.

with the experiment. The observer is then entangled with the experiment. This illustrates that taking into account both possible states with their probability will lead to correct predictions, while assuming that the cat is in only one of the two states will likely give incorrect results.

Dirac (1902–1984) starts from the equation of a particle in relativistic classical mechanics, $W^2/c^2 - p^2 - m^2c^2 = 0$, to obtain a wave equation in quantum mechanics. The left-hand side operates on the wave function ψ, and W and p are replaced by the operators $ih\partial/\partial t$ and $-ih\partial/\partial r$. This gives $[W/c - \alpha_r p - \alpha_o mc]\psi = 0$. These new variables α_r and α_o give rise to the spin of the electron. "From the general principles of quantum mechanics one can easily deduce that these variables α give the electron a spin angular momentum of half a quantum and a magnetic moment of one Bohr magneton in the reverse direction to the angular momentum." Dirac's theory also proves the existence of positive electrons having the same mass and opposite charge to the well-known negative electrons. All those points are confirmed experimentally. He also adds in his Nobel lecture that "If we accept the view of complete symmetry between positive and negative electric charge so far as concerns the fundamental laws of Nature, we must regard it rather as an accident that the Earth (and presumably the whole solar system), contains a preponderance of negative electrons and positive protons. It is quite possible that for some of the stars it is the other way about, these stars being built up mainly of positrons and negative protons. In fact, there may be half stars of each kind. The two kinds of stars would both show exactly the same spectra, and there would be no way of distinguishing them by present astronomical methods."

Schrödinger and Dirac are awarded the Nobel Prize in 1933 for "the discovery of new productive forms of atomic theory".

Finally, Pauli (1900–1958) receives the Nobel Prize in 1945 "for the discovery of the Exclusion Principle, also called the Pauli Principle". As we have seen, Bohr's model of the atom postulates that electrons move in stationary orbits around the nucleus. To better characterize each of the electrons of an atom, quantum numbers are assigned and correspond to distinct energy levels and spatial extents: principal quantum number (n), the azimuthal (also known as orbital angular momentum) quantum number (l), the magnetic quantum number (m) (Figure 3.16). In 1925, Pauli introduces the spin quantum number (m_s), which takes values of +1/2 or –1/2, and formulates the Pauli principle: no two electrons in an atom can have identical sets of quantum numbers. Let us add that the Pauli principle also stands for protons and neutrons.

FROM THE COMPUTATION OF QUANTUM PROPERTIES TO QUANTUM COMPUTING

"There is an oral tradition that, shortly after Schrodinger's equation for the electronic wave-function ψ had been put forward and spectacularly validated for simple small systems like He and H_2, P. M. Dirac declared that chemistry had come to an end – its content was entirely contained in that powerful equation. Too bad, he is said to have added, that in almost all cases, this equation was far too complex to allow solution." (Kohn) So, the idea in the following years is to introduce approximations. Let us start

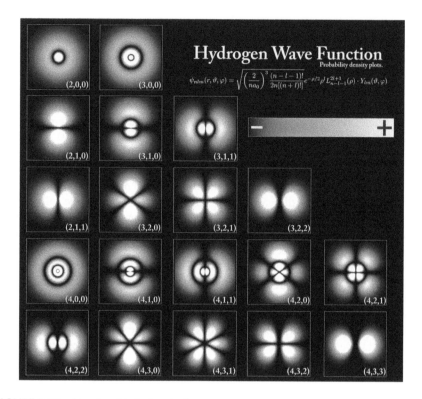

FIGURE 3.16   Atomic orbitals for a hydrogen atom.

with the Born–Oppenheimer approximation. It allows the separation of the motion of nuclei from the motion of electrons. Since the mass of an electron is much smaller than the mass of a nucleus, the nucleus moves very slowly when compared to an electron and, thus, its motion can be neglected in the Schrödinger equation. This means that the atomic nucleus can be placed at different stationary positions and that the electronic wave function can then be calculated as a function of the positions of the nucleus. In other words, the wave function is evaluated for a series of "snapshots" for different positions of the atomic nucleus, even though its motion is neglected! Then, different methods can be used to determine the electronic structure. One of the first methods is introduced in 1927 by Hartree to calculate approximate wavefunctions and energies for atoms and ions. He calls it the self-consistent field (SCF) method (now known as the Hartree–Fock (HF) method). The idea is to start by assuming that electrons do not interact directly with each other. Mathematically, it amounts to approximating the wave function of a many (N-body) particle system (for example when $N > 1$ electron, i.e. $N = 2, 3, \ldots$) as the "product" of orthogonal wave functions of the individual particles. This product is known as the Hartree product. Nevertheless, it does not satisfy the Pauli exclusion principle (two electrons cannot occupy the same quantum state). For example, for He, we have two electrons in the electron configuration $1s^2$. There is thus one electron in the quantum state ($n = 1$, $l = 0$, $m_l = 0$, $m_s = 1/2$) and one electron in the quantum state ($n = 1$, $l = 0$, $m_l = 0$, $m_s = -1/2$). This means that the "approximate" wave function needs to be antisymmetric. This is the reason why Slater and Fock use the "Slater determinant", in lieu of the Hartree product. By doing so, they have a product, but also an antisymmetric "approximate" wave function. Then, finally, to account for the interactions of an electron with its surroundings, the Hamiltonian for each electron also includes, in addition to its kinetic energy, a mean-field Coulomb potential. This considerably simplifies the problem and yields one-electron Schrödinger equations. Applying the operator $\hat{H}$ to this "approximate" wave function and calculating the energy expectation $E = <\psi|H|\psi>$ yields an upper estimate for the ground state energy of the system. Then, using a variational principle, one can find the minimum for the energy E corresponding to this ψ by seeking the optimal set of single-particle states (φ_k) that makes $<\psi|H|\psi>$ stationary under infinitesimal changes of φ_k. Each ψ_i is indeed an expansion on a basis set: $\psi_i = \sum_k^M c_{ik} \varphi_k$. This means that, if M basis functions φ_k are known, one can find the appropriate c_{ik} that minimize the energy, giving the wave function and the energy of the ground state (see Figure 3.17). In general, φ_k are localized Gaussians in finite systems and a basis of plane waves in periodic systems. However, even with these approximations, the equations can become very complicated and impossible to solve by hand. At the beginning of the 1960s, the rise of computers changes everything. Pople (1925–2004) decides to create a computer program to calculate, among other things, molecular structure, ground and excited states and reaction mechanisms. It constantly evolves and uses more and more refined approximations. It is still widely used around the world! In 1998, Pople receives the Nobel Prize "for his development of computational methods in quantum mechanics".

However, a method such as Hartree–Fock does not fully take into account the "exchange" interactions or the electrostatic repulsion between electrons, resulting in incorrect predictions even for simple metals. A scheme is then developed to

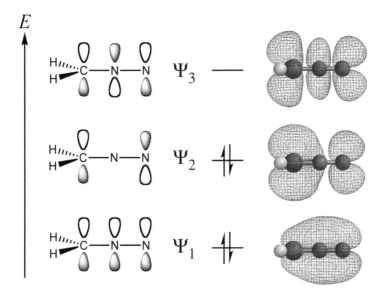

FIGURE 3.17 Hartree–Fock results for the diazomethane π system.

incorporate the effects of both exchange and correlation: the density functional theory. In two landmark articles in 1964 (Hohenberg and Kohn) and 1965 (Kohn and Sham), Kohn shows that another type of approximation can be made, through the use of electron density. Indeed, he demonstrates that there is a one-to-one correspondence between the energy of a quantum mechanical system and its electron density! Since the latter is a function of only three coordinates (x,y,z), it is therefore much more convenient to handle rather than complicated wave functions, which depend on the positions of all electrons. This new method is based on two theorems. The first one states that the ground state properties of a many-electron system only depend on the electronic density $\rho(x,y,z)$. The second states that the correct ground state density for a system is the one that minimizes the total energy through the functional $E[\rho(x,y,z)]$. Let us recall that in that case, one begins by calculating ρ from (x,y,z), giving $\rho(x,y,z)$. Then $\rho(x,y,z)$ becomes the input of a function E that will give the famous $E[\rho(x,y,z)]$. The idea is then to determine the density functional $E[\rho]$. Since $H_{el} = T_e + V_{eN} + V_{ee}$ (T_e: kinetic energy, V_{eN}: interaction nucleus–electron and V_{ee}: interaction electron–electron), it can be rewritten in the DFT formalism as

$$E[\rho] = T[\rho] + E_{eN}[\rho] + E_{ee}[\rho]$$

The Kohn–Sham equations become

$$E_{KS}[\rho] = T_s[\rho] + E_{eN}[\rho] + J[\rho] + E_{xc}[\rho]$$

with

$$E_{xc}[\rho] = (T[\rho] - T_s[\rho]) + (E_{ee}[\rho] - J[\rho])$$

and

$$\rho(r) = \sum_{i=1}^{N} |\varphi_i(r)|^2$$

and

$$T_s[\rho] = \sum_{i=1}^{N} < \varphi_i \,|-1/2\, \nabla^2|\, \varphi_i >$$

As we can see, one can compute every piece of the Kohn–Sham density functional theory (KS-DFT) energy exactly except for the "exchange-correlation" term, $E_{xc}[\rho]$. Hence, various approximate exchange-correlation functions have been developed over the past 50 years. DFT has become a versatile computational tool with many applications in chemistry, physics and materials science. The computational cost is comparable to HF but the accuracy of the calculations is often better since correlation is accounted for through the "exchange-correlation" functional. Due to its simplicity, it can be applied to larger molecules and more complex materials than wave function-based methods. For instance, DFT has shown to be very instrumental for the understanding of enzyme catalysis, photosynthesis, computer-aided drug design and the discovery of new materials. Kohn receives in 1998 the Nobel Prize in Chemistry "for his development of the density-functional theory". In recent years, multiscale computations that combine classical and quantum mechanics have shed light on chemical reactions and complex biochemical processes including drug-protein interactions. In 2013, Karplus, Levitt and Warshel receive the Nobel Prize in Chemistry "for the development of multiscale models for complex chemical systems".

Nowadays, quantum technologies are fast developing. The sheer amount of data produced every day challenges the capability of conventional computers. Indeed, every day, 2.5 exabytes (2.5×10^{18} bytes) are produced! As a result, scientists and tech companies have started to focus on developing the next generation of computers: quantum computers. Quantum computers can considerably decrease the time it takes to complete calculations and solve extremely complex problems. For example, powerful quantum processors could handle huge data sets (Big Data) and, combined with artificial intelligence, analyze them in record time! This could be done by taking advantage of quantum properties such as superposition to store information using qubits (quantum bits). As we know, for an electron, there are two possible quantum states: spin up $|0>$ and spin down $|1>$. When there is superposition, we have a complex combination of both states, and the wave function can be represented as $|\psi> = \alpha_0 |0> + \alpha_1 |1>$ with $0 < |\alpha_0|^2 < 1$ being the probability of being in the spin up state and $0 < |\alpha_1|^2 < 1$ in the spin down state. This means that while there are only two possible bits in conventional computers to store the information (the famous binary code 0 and 1), there are many possible states for a qubit, depending on the values of α_0 and α_1 (Figure 3.18). However, in practice, storing a quantum state – i.e. particles in superposition – is very difficult. Any interaction with the surroundings, such as communication with the other qubits in a quantum computer, will disrupt superposition and cause errors. This is the reason why quantum computers are shielded electromagnetically and cooled down to almost absolute zero. Another principle, known as quantum entanglement, has recently emerged as a promising candidate for novel applications such as networking, quantum cryptography and quantum key distribution. The idea is that when there is a pair or group of particles, acting on one of them instantly influences the others, even if they are separated by a large distance. This phenomenon is sometimes described as quantum teleportation.

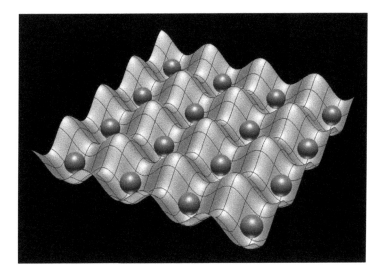

FIGURE 3.18 Qubits in neutral-atom quantum processors.

4 Introduction to Electrochemistry
Electron Transfer

Dieser Idealismus ist in diesem Widerspruche, weil er den abstrakten Begriff der Vernunft als das Wahre behauptet; daher ihm unmittelbar ebensosehr die Realität als eine solche entsteht, welche vielmehr nicht die Realität der Vernunft ist, während die Vernunft zugleich alle Realität sein sollte; diese bleibt ein unruhiges Suchen, welches in dem Suchen selbst die Befriedigung des Findens für schlechthin unmöglich erklärt.—So inkonsequent aber ist die wirkliche Vernunft nicht; sondern nur erst die Gewißheit, alle Realität zu sein, ist sie in diesem Begriffe sich bewußt als Gewißheit, als Ich noch nicht die Realität in Wahrheit zu sein, und ist getrieben, ihre Gewißheit zur Wahrheit zu erheben, und das leere Mein zu erfüllen.

Hegel, *"Phänomenologie des Geistes"*, 1807

This idealism falls into such a contradiction because it asserts the abstract Notion of Reason to be the Truth; consequently, Reality comes directly before it just as much in a form which is not strictly the Reality of Reason at all, whereas Reason all the while is intended to be all Reality. Reason remains, in this case, a restless search, which in its very process of seeking declares that it is utterly impossible to have the satisfaction of finding. But actual concrete Reason is not so inconsequent as this. Being at first merely the certainty that it is all Reality, it is in this notion well aware that qua certainty qua ego it is not yet in Truth all Reality; and thus Reason is driven on to raise its formal certainty into actual Truth, and give concrete filling to the empty "mine".

Hegel, "The Phenomenology of Mind", 1807

ELECTRICITY AND THE "ELECTRIC FLUID"

Interestingly, we find the root of the word "electricity" in the ancient Greek name: Ηλέκτρα (Electra) which translates into "amber", "shining". In Greek mythology, Electra is one of Agamemnon's daughters, the famous King of Mycenae, who heads the army in charge of rescuing Helen from Paris, the son of Priam, king of Troy (see Figure 4.1). It is the famous Trojan war! We find some information about the Trojan war in Homer's "Odyssey" (675–725 BC). Indeed, this poem relates the story of Odysseus, who after taking part to the Trojan war (he was among the soldiers hidden in the famous Trojan Horse), spends ten years to return home to Ithaca! Homer's second

FIGURE 4.1 Scenes from the Trojan war.

most known epic is the "Iliad", which comes from Ilion (the other name of Troy!). It tells the events occurring over the last year of the siege of Troy and, especially, the meddling of gods in mortals' affairs. Let us add that Virgil, the greatest Roman poet, later draws inspiration from Homer's "*Odyssey*" and "*Iliad*" to create his most famous epic: "*Aeneid*". It is the story of Aeneas, a Trojan, who after the war leaves Troy and embarks on a journey that takes him through adventures and wars to finally build a new city: the cradle of the Roman Empire. There is also another mention of Electra in Greek mythology. Indeed, she is one of the seven daughters of Atlas and Pleione, who are known as the Pleiades. According to a myth, they were turned into stars by Zeus to escape Orion. Nowadays, in astronomy, the Pleiades is the name of a very bright open star cluster in the constellation Taurus known as the "Seven Sisters" or "M45". This cluster is located 4.2×10^{15} km from Earth but can be seen with the naked eye! Finally, another myth tells the fate of Phaeton, son of the Sun god (Helios). After borrowing the chariot of the sun from his father, he loses control of the horses and comes too close to the Earth, beginning to scorch it. To avoid humankind's destruction, Zeus strikes Phaeton with lightning, who falls dead into a river. His sisters, known as the Heliades, unable to stop crying over his death are changed into trees along the river by Zeus. Their tears turn into amber. Virgil writes about it in his poem "*Eclogues*", also called "*Bucolics*". Later, Ovid tells this story in his "*Metamorphoses*":

> 'parce, precor, mater,' quaecumque est saucia, clamat,
> 'parce, precor: nostrum laceratur in arbore corpus
> iamque vale' – cortex in verba novissima venit.
> inde fluunt lacrimae, stillataque sole rigescunt
> de ramis electra novis, quae lucidus amnis
> excipit et nuribus mittit gestanda Latinis.

Oh, spare me, mother; spare, I beg you. 'Tis my body that you are tearing in the tree. And now farewell' – the bark closed over her latest words. Still their tears flow on, and these tears, hardened into amber by the sun, drop down from the new-made trees. The clear river receives them and bears them onward, one day to be worn by the brides of Rome.

This is the reason why, for centuries after, amber is associated with eternal youth and as a talisman guaranteeing the protection of gods.

And amber also lies at the origin of electricity. With the birth of Greek philosophy between 600 BC and 400 BC, the principles of static electricity are discovered. It begins with the study of the properties of amber (the fossilized resin from trees). Indeed, when amber is rubbed by fabric, it attracts small bodies such as wool, hair and straw. The same is true for lodestone or magnetite that can attract objects, but also repel others. De Maricourt, also known as Peregrinus, explains how this strange behavior can be used to make a compass (see Figure 4.2) in "*Epistola de Magnete*" (1269). "By this very instrument, you may direct your course to any cities and islands, to whatever other places, wherever you wish to by land or sea, with the longitude and latitude of these places always known to you." Such devices allow for the discovery of America by Columbus in 1492 and the first circumnavigation of the Earth by Magellan between 1519 and 1522.

In "*De Magnete*" (1600), Gilbert (1544–1603), the first "electrician", introduces several definitions. He is Queen Elizabeth I's and then King James I's physician. He is the first to understand how a compass really works. He gives the name of "pole to the extremities of the magnetic needle pointing to the poles of the earth, calling

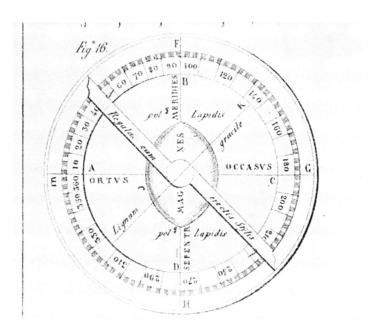

FIGURE 4.2   Mariner's compass (13th century).

south pole the extremity that pointed toward the north, and north pole the extremity pointing towards the south". He applies the term "magnetic" to all bodies which are acted upon by loadstones and magnets, and he finds that all such bodies contain iron in one state or another! He considers the phenomenon of electricity as having a considerable resemblance to magnetism, although he points out the differences between the two. According to him, "Electrics are bodies that attract in the same way as amber. An excited magnetic body is one (such as iron or steel) that acquires magnetism from a loadstone or natural magnet". He has the intuition that attraction is connected to the electric or magnetic nature of objects. He understands that: "very many electric bodies (as precious stones, etc.) do not attract at all unless they are first rubbed" while "the loadstone, through it is susceptible of a very high polish, has not the electric attraction". According to him, "Electrical movements come from the materia (matter), but magnetic from the prime forma (form); and these two differ widely from each other." He concludes "A loadstone attracts only magnetic bodies; electrics attract everything [...] A loadstone lifts great weights; [...] Electrics attract only light weights." Because of his work on magnetism, the unit for magnetomotive force (property that gives rise to magnetic fields) is called the Gilbert (Gb), although it is a rather old unit of measurement.

Unlike Gilbert, who thinks that the Earth is composed of iron and thus is magnetic, von Guericke (1602–1686) believes that the Earth is an electric body. To demonstrate his point, he carries out many experiments reported in *"Experimenta Nova"* (1672). He starts by building the first electrostatic generator in 1663. He places a ball of sulfur on an axle and spins the ball, while a cloth rubs the ball surface. The ball gives off sparks! And also attracts light pieces of straw. He also discovers the phenomenon of electroluminescence. In 1705, one of Newton's assistants, Hauksbee (1660–1713), builds on von Guericke's experiment, but, this time, using a vacuum pump and a glass globe. He presents his experiment before the Royal Society. He starts with a glass globe, from which air can be pumped in and out, on a rotating spindle. He puts his hands on the surface of the globe and observes a very bright light when air is pumped out of the globe. Moreover, "The light appear'd of a curious Purple colour". He also notices that "a light would appear to ftick to the Fingers, notwithftanding they did not touch the Glafs [...] and my Neckcloth at the fame time, at an inch or 2 diftant from it, appear'd of the colour of Fire, without any communication of Light from the Globe" giving birth to new phenomena created by "electric fire".

The nature of this "electric fire" draws considerable attention during the 18th century. In 1733, Du Fay (1698–1739) proposes the existence of two types of electricity: vitreous and resinous. Indeed, the nature of electricity depends on the material that is rubbed. For example, when glass is rubbed, it creates a vitreous electric fluid whereas rubbing resin (amber) or wax produces a resinous electric fluid. He also notices that two substances electrified with the same fluid repel one another (for example, two objects rubbed against amber repel each other). On the other hand, two substances electrified with different fluids attract each other (an object rubbed against amber and another rubbed against glass attract each other). At about the same time, another theory is proposed by Franklin (1706–1790). Indeed, he explains these phenomena with a single electrical fluid or "electrical fire". His idea is that all materials are composed of this electric fluid, and that, through friction, this fluid is simply transferred from one body to another. A body therefore bears a positive charge (or is "over-electrified") when it receives an excess of electrical fluid (the quantity of electrical fluid after the transfer is greater than the quantity it had before the transfer). On the other hand, when a body has a lack of electrical fluid, it then bears a negative charge and is called "under-electrified". He also postulates that "electricity flows from an over-electrified body (positive) to an under-electrified body (negative)"; this will be proved incorrect later on with the discovery of the electron. Nevertheless, nowadays, we use the same convention as Franklin's, that is the direction of the electric current goes from positive to negative. Another of Franklin's discoveries is the existence of natural electricity (not produced by friction). In his famous 1752 experiment with a kite, he shows that he can extract electricity from a cloud, meaning that lightning is of an electric nature. According to Priestley "The kite being raised, a considerable time elapsed before there was any appearance of its being electrified. One very promising cloud had passed over it without any effect; when at length, just as he was beginning to despair of his contrivance, he observed some loose threads of the hempen string to stand erect, and to avoid one another, just as if they had been suspended on a common conductor. Stuck with this promising appearance, he immediately presented his knuckle to the key, and (let the reader judge of the exquisite pleasure he must have felt at that moment) the discovery was complete. He perceived a very evident electric spark." Let us add that Franklin is one of the Founding Fathers and that he helps write the Declaration of Independence. His numerous scientific discoveries considerably improve the standard of living in the USA during the 18th century thanks to, for instance, the invention of lightning rods, bifocals and stoves as well as through the foundation of the first public library and the first company of firefighters in Philadelphia.

However, the origin of the electric fluid remains elusive. Moreover, a question arises as two negative charges repel each other, meaning that there should exist a repulsive force, but at this time, only the attractive, Newtonian, gravitational force is known.

ELECTRICITY AND MAGNETISM: TOWARD ELECTROMAGNETISM

Coulomb (1736–1806) finds the answer by showing that the effects of electricity can be understood as a "reaction force". Indeed, he builds the famous "Coulomb balance" (see Figure 4.3) which "measures very accurately the state and the electric force of a body regardless of how weak the degree of electricity is". In his "Premier Mémoire sur l'éléctricité et le magnétisme", in 1785, he explains the operation of this apparatus and demonstrates that the electric force is similar to the gravitational force. More specifically, he defines the following law: "the repulsive force of two small globes with the same nature of electricity is inversely proportional to the square of the distance between the centers of the two globes". He has more trouble measuring the attraction force between electric bodies of opposite charges, since the two bodies jump and stick to each other when they get too close. Refining his experiment, Coulomb finds that "we have arrived here by a method absolutely different from the first [...] to conclude that the attraction of the electric fluid 'positive' for the electric

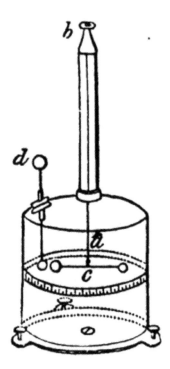

FIGURE 4.3 Coulomb electrometer.

fluid 'negative' is as the inverse square of the distance". However, most of his contemporaries refute his conclusions. Finally, thanks to his French colleagues, his work will be honored by giving his name to the law he discovered (Coulomb's law) and, later, to the unit of the electric charge ("Coulomb", noted C). One Coulomb (1 C) is the amount of electricity traveling through a conductor for a current of 1 Ampere and for 1 second. It is equivalent to 6.24×10^{18} elementary charges. For a positive charge of 1 C, it represents 6.24×10^{18} protons. Alternatively, for a negative charge of –1 C, it represents 6.24×10^{18} electrons!

Until 1820, electricity and magnetism are assumed to arise from two distinct forces. During an experiment, Ørsted (1777–1851) observes the following. When he runs an electric current through a wire and approaches a compass, he notices that the needle of the compass moves! This means that electricity and magnetism are connected. Over the next few months, he carries out more experiments, gathered in his paper "Experiments on the Effect of a Current of Electricity on the Magnetic Needle". In addition, he finds that "all the effects on the north pole above-mentioned are easily understood by supposing that negative electricity moves in a spiral line bent towards the right, and propels the north pole, but does not act on the south pole. The effects on the south pole are explained in a similar manner, if we ascribe to positive electricity a contrary motion and power of acting on the south pole, but not upon the north." In other words, he finds that reversing the electric current moves the needle in the opposite direction. This means that an electric current can produce a magnetic field, a phenomenon he calls "electric conflict". In September 1820, Arago introduces Ørsted's experiment to French scientists, which prompts Ampère to study electromagnetism. The British Royal Society awards Ørsted the 1820 Copley Medal, the greatest prize in science at the time, "For his Electro-magnetic Discoveries". Previous prize winners include Franklin, Cavendish, Priestley, Cook, Herschel, Thompson (Count Rumford), Volta and Davy.

From 1820 on, Ampère (1775–1836) (see Figure 4.4) works on Ørsted's experiment and discovers the famous "*règle du bonhomme d'Ampère*", also known as the "*Ampère rule*" or "right-hand rule". Specifically, it allows the prediction of the direction in which the needle of a compass is deflected when an electric current flows in its vicinity. Let us imagine a man with the electric current running from his feet to his head and with his eyes directed towards the needle. When he extends his left arm, it indicates the direction of the north of the needle! Also, when Ampère uses a long current-carrying electric wire and a magnet, he observes that when he moves the magnet along a circle around the wire, the direction of the north pole changes, mapping out the circular loop! The right-hand rule shows that, at any time, one can deduce the direction of the north pole. Indeed, on a right-hand grip position, if the thumb is along the direction of the electric current, then the other fingers show the direction of the north pole! Ampère goes further and finds that, if an electric current can act on a magnet, it means that there must be electric currents within the magnet. In other words, the action of an electric current on a magnet is simply the action of this electric current on electric currents contained within the magnet! Around 1822, Ampère

FIGURE 4.4 André-Marie Ampère (1775–1836).

carries out more experiments and shows that, when he moves a current-carrying test wire close to another current-carrying wire, there is a force that takes place between the two. Moreover, he observes that this force is proportional to the length of the test wire. Also, he notes that when the two currents flow in the same direction, the two wires attract one another! On the other hand, if he reverses the current in one of the wires, then the two wires repeal each other! Nowadays, Ampère's law allows for the calculation of the magnitude of a magnetic field $|\mathbf{B}|$ or B created by a current-carrying wire. If the current has an intensity I, Ampère's law states that $\int_c \mathbf{B} \cdot \mathbf{ds} = \mu_0 I$ where the integral (circulation) is taken over the circumference of the circular loop of radius r and μ_0 is a constant. This means that, in this case, $B \times 2\pi r = \mu_0 I$ or $B = \mu_0 I/(2\pi r)$, which, in the case of a long current-carrying wire, is simply proportional to the ratio of the current flowing through to the circumference of the circular loop of radius r. Unfortunately, this discovery is rejected by the scientists of the time. It is only 60 years later that the discovery of electrons confirms this theory of magnetic fields! Nowadays, the Ampere (A) is the unit of measure of the intensity of an electric current.

However, there are no "magnetic charges" (also called magnetic monopoles), analogous to electric charges. Indeed, magnetic fields are generated by a dipole. It is during his explorations in South America that von Humboldt (1769–1859) notices strange things happening with his compass. Indeed, he observes irregular disturbances that he names "magnetic storms". To better understand these phenomena, Gauss (1777–1855) starts to study the Earth's magnetic field. To do so, he uses a method similar to that used in celestial mechanics to study the effect of gravity. It consists of defining the Earth as a sphere, and the magnetic field generated by the core of Earth as a sum of "north–south" multipoles: "two-pole" (like a magnet) whose strength decreases as $1/r^3$ + "four-pole" whose strength decreases $1/r^4$ + "eight-pole" decreasing as $1/r^5$ + ... Von Humboldt's and Gauss' results show that the dipole part is the most important! It also provides a new map of the magnetic field on Earth and an answer to von Humboldt's observations of magnetic storms! Gauss also demonstrates that magnetic monopoles do not exist. Indeed, one proof can be found in the now so-called Gauss' law of magnetism. It states that the magnetic flux through any closed surface is equal to zero. In other words, the quantity of magnetic field passing through a closed surface is zero! If we look at its correspondence in electricity, the Gauss' law of electricity states that the electric flux through any closed surface is proportional to the net electric charge enclosed by that surface. Comparing these two laws, it can be seen that there is no "magnetic charge" (since it is equal to zero), and thus no existence of magnetic monopoles! It also means that the basic entity for magnetism is the magnetic dipole!

Finally, in 1860, Maxwell (1831–1879) brings together all known laws on electricity and magnetism to form his famous electromagnetic theory. Let us add that, in 1885, Heaviside will summarize it in only four equations! The first law (Gauss' law) stipulates that electric charges produce an electric field ($\nabla \cdot \mathbf{D} = \rho$). The second law (Faraday's law) states that a changing magnetic field gives rise to an electric field ($\nabla \times \mathbf{E} = -\partial \mathbf{B}/\partial t$). The third law (Ampère's law) quantifies the electric current created by a magnetic field ($\nabla \times \mathbf{H} = \mathbf{J}$). And the fourth law (Gauss' law for a magnetic system) demonstrates that no magnetic charges can exist ($\nabla \cdot \mathbf{B} = 0$). Moreover, Maxwell proposes that, even in an insulator, electric charges are displaced by a small distance. This displacement, in turn, creates a current $\partial \mathbf{D}/\partial t$. He then reformulates Ampère's law as $\nabla \times \mathbf{H} = \mathbf{J} + \partial \mathbf{D}/\partial t$. Together with Faraday's law $\nabla \times \mathbf{E} = -\partial \mathbf{B}/\partial t$, it shows that electricity and magnetism are coupled! Differently put, any disturbance in the electric and magnetic fields will travel out together in space at the speed of light as an electromagnetic wave!

VOLTA'S PILE AND THE BIRTH OF ELECTROCHEMISTRY

The formation of sparks captures the popular imagination at the end of the 17th century. More and more experiments are designed to produce even more powerful sparks. An idea among many others is to draw electric fire from water. Two scientists, von Kleist and van Musschenbroeck, follow this route and discover that it is possible to store electricity. In 1745, Van Musschenbroeck, who works in Leiden, creates the famous Leiden jar (see Figure 4.5)! He attempts to electrify water in a

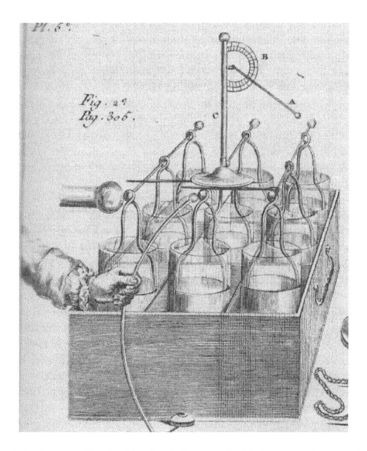

FIGURE 4.5   An assembly of Leiden Jars to form an electric battery (18th century).

glass flask. To do this, he uses an electrostatic machine and connect it with a chain to a metal rod dipped in the water flask. When he touches the metal rod, he feels an electric shock! This shows that the water flask can accumulate and store electricity! He also establishes that, in order to produce more powerful sparks, one only needs to increase the capacity (volume) of the jar! In 1785, van Marum (1750–1837) creates a "monster" frictional electrostatic machine using, among other things, 100 Leyden jars of 13 inches diameter each and 2 feet high. This machine is now part of the collection of the Teylers Museum in the Netherlands. It can deliver 24-inch sparks, corresponding to 720,000 V! Van Marum also notices a particular odor that he calls "smell of the electrical matter". It is not until 1840 that Schönbein smells the same scent and decides to call it ozone from Greek ὄζειν = ozein meaning odor.

At the end of the 18th century, the nature of electricity remains an unresolved issue. However, in 1786, Galvani (1737–1798) makes a new discovery. He finds that, when he discharges a Leyden jar in the leg muscle of a dead frog, the muscle contracts to his great amazement! To confirm this finding, he carries out experiments in which he uses an electrostatic machine that produces sparks, as well as lightning. Every single time, the muscle contracts! According to Galvani, there is thus a new form of electricity he calls "animal electricity". In 1803, his nephew (Aldini) goes even further and tries this technique on the corpse of a recently executed criminal. By applying electricity on the face of the criminal, some muscles contract, making him look as he were alive and trying to speak! These experiments of "galvanized corpses" become very famous around Europe. One illustration of this craze is the novel "Frankenstein" by Shelley in 1818. This fantastic story is based on Dr. Frankenstein's experiment, in which, using galvanism, he tries to bring back to life a corpse, made of different dead body parts stitched up together. Let us add that there are less morbid, and even lifesaving, applications of galvanism. For instance, nowadays, the defibrillation machine is used after a cardiac arrest to restart someone's heart with an electric shock!

Another groundbreaking discovery emerges from the famous controversy between Galvani and Volta over the concept of "animal electricity". Indeed, Galvani and Volta (1745–1827) are friends. The first is a physician and the other a physicist. At first, Volta is extremely impressed by Galvani's findings. However, with time, Volta begins to think that electricity comes from the metals used during Galvani's experiments (scalpels, hooks). Indeed, he detects some electricity by placing different metals on his tongue. This is the reason why Galvani carries out additional experiments to show that muscle contraction occurs even in the absence of any metal. Meanwhile, Volta continues his experiments on metals and demonstrates that the use of two different types of metals can produce electricity! In 1800, he builds the first electric battery known as "Volta's pile" or "voltaic column" (see Figure 4.6). He presents his apparatus in "On the Electricity Excited by the Mere Contact of Conducting Substances of Different Kinds" at the Royal Society of London. Specifically, "it consists of a long series of an alternate succession of three conducting substances, either copper, tin and water; or, what is much preferable, silver, zinc, and a solution of any neutral or alkaline salt". In addition, "The mode of combining these substances consists in placing horizontally, first, a plate or disk of silver (half-a-crown, for instance) next a plate of zinc of the same dimensions; and, lastly, a similar piece of a spongy

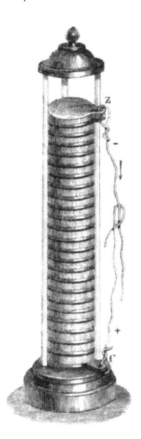

FIGURE 4.6   Volta's pile.

matter, such as pasteboard or leather, fully impregnated with the saline solution. This set of three-fold layers is to be repeated thirty or forty times, forming thus what the author calls his columnar machine." He realizes that, when he puts one finger at the bottom of the column and another touching the top of the column, he feels a succession of small electric shocks! Even for only 20 pairs of metallic plates! This is the birth of the first battery! "With a column of about sixty pairs of plates, shocks have been felt as high as the shoulder." In other words, the higher the stack, the greater the shock. Volta's "columnar machine" is thus a convenient and dependable source of electricity that quickly leads to the emergence of a new field, the phenomenon of electrolysis. In 1836, the Daniell cell revolutionizes the world of electricity giving rise to our modern batteries.

Within six weeks of Volta's announcement, Nicholson and Carlisle perform the first water electrolysis. They start by redoing Volta's experiments and connect a voltaic battery to an electroscope, a device that measures the quantity of electricity produced by the battery. To ensure that their measurements are accurate, they add a drop of water on the top plate of the battery and dip the wire of the electroscope in the water. To their great surprise, bubbles of gas quickly form in the water droplet! This gas is what we now know as hydrogen gas. Then, Nicholson and

Carlisle decide to carry out a different experiment. They fill a test tube with water and dip two wires in the water, one connected to the top plate of the battery and the other to the bottom plate. They observe bubbles around each of the two wires! Even if the two wires are separated by almost two inches! In other words, wires connected to the two poles of the battery give rise to two gases, later identified as the two components of water, hydrogen and oxygen! This shows that galvanizing water triggers a chemical reaction, breaking water into two gases. This phenomenon quickly becomes known as electrolysis (from the Greek λύσις: *lysis* = dissolution) and marks the beginning of a new field: electrochemistry! Nicholson and Carlisle publish their findings in Nicholson's Journal in 1800. Initially received with skepticism, Nicholson and Carlisle's work becomes rapidly accepted and quickly confirms Lavoisier's new ideas on chemistry. Electrochemistry also yields rapid advances. Indeed, in the following years, Davy uses electrolysis and discovers a series of new elements!

Another major contribution comes from Berzelius (1779–1848). While attending medical school at Uppsala, he becomes interested in Volta's battery. Indeed, he wants to understand the effects of electric shocks on patients, known as galvanotherapy at the time. However, he finds that it has little effect. In 1803, he introduces the concept of "dualism" in order to explain how Volta's battery decomposes chemicals like water. According to him, chemicals are split into pairs of electrically opposite constituents. For instance, in the case of water, electrolysis gives an electropositive hydrogen and an electronegative oxygen! He also discovers that salts are made of electronegative acids and electropositive bases, meaning that according to Berzelius and his theory on dualism, all chemical compounds are composed of two electrically opposing constituents: the acidic, or electronegative, component and the basic, or electropositive, component! Nevertheless, the idea of dualism faces a major challenge with the emergence of a new discipline known as organic chemistry. Let us note there is also a secret society at Yale University named after Berzelius. Founded in 1848, 'BZ' is the third oldest society at Yale and the oldest from the Sheffield Scientific School (formerly its sciences and engineering college). The "Big Three" societies at Yale are: "Skull and Bones" founded in 1832 (famous alumni: George H. W. Bush, George W. Bush, John Kerry), "Scroll and Key" founded in 1842 (famous alumni: Rockefeller and Leonard Case Jr., Cole Porter) and "Wolf's Head Society" (The Third Society) founded in 1883 (famous alumni: William Clay Ford, A. Conger Goodyear, William Wrigley III).

FARADAY AND THE LAWS OF ELECTROLYSIS

Davy (1778–1829) is at the center of many discoveries in chemistry at the end of the 18th and beginning of the 19th century. Indeed, he starts to study pneumatics and especially the therapeutic use of gases. In 1799, he discovers the famous "laughing gas" (nitrous oxide N_2O) and recommends it as an anesthetic for minor surgery, but it soon becomes an attraction in social gatherings! In fact, Davy also gives lectures at the Royal Society which enthrall the public and draw large audiences. The contemporary writer, Dibdin, witnesses the lecture and recalls: "The whole had the character of a noonday opera house. There stood Davy, every Saturday morning, as the magician of nature – as one, to whom the hidden properties of the earth were developed by some Egerian priestess in her secret recess. [...] The tremendous force of such an agency struck the learned with delight, and the unlearned with mingled rapture and astonishment; and the theater or lecture room rung with applause as the mighty master made his retreating obeisance." In his goal to apply science for good, he also invents a new lamp for coal miners in which the flame is enclosed within the lamp, and thus does not ignite the gases present in the mine. Before the Davy lamp, miners only used torches! Volta's battery discovery in 1800 prompts him to begin studying the effects of electricity on chemicals. He presents his experiments in a lecture entitled "On Some Chemical Agencies of Electricity" in 1806. In particular, he uses a gigantic voltaic battery composed of 500 copper and zinc plates, that transforms potash into potassium! In other words, a colorless stone turns into a shiny silvery metal! He also demonstrates that "the hardest metals melted like wax beneath its operation". Using this device, he also discovers sodium (Na), magnesium (Mg), calcium (Ca), strontium (Sr) and barium (Ba), a new family of chemical compounds known as today alkaline-earth metals! Davy also notes that "hydrogen, the alkaline substances, the metals, and certain metallic oxides, are attracted by negatively electrified metallic surfaces, and repealed by positively electrified metallic surfaces; and contrariwise, that oxygen and acid substances are attracted by positively electrified metallic surfaces, and repealed by negatively electrified metallic surfaces; and these attractive and repulsive forces are sufficiently energetic to destroy or suspend the usual operation of elective affinity." For his research, Davy receives numerous awards and honors, among them the Copley Award. He is also knighted in 1812 and becomes a baronet in 1818.

After attending Davy's lectures, a young bookbinder, Faraday (1791–1867) (see Figure 4.7), writes up his notes and sends them to Davy, while asking him for a job. He becomes Davy's assistant in 1812. From there, Faraday's scientific career begins; he will become one of the most influential scientists of the 19th century! He starts with the study of electromagnetism. First, he discovers that, by reversing the Oersted experiment, one can build an electric motor! To do so, he puts a magnet in a bath of mercury and dips one of the ends of a wire in the bath and the other end of the wire is fixed at the bottom of the mercury bath. The wire is connected to a battery. When the current flows in the wire, he notices that the wire spins around the magnet in circles. It stops when the current is switched off. This means that the current flowing through the wire generates a magnetic field (as discovered by Ampère a year before) and that this magnetic field, in turn, interacts with the magnet in the bath. This electromagnetic interaction creates the motion of the wire! This will later lead to the design of electric motors. In 1831, he discovers the phenomenon of "electromagnetic induction". This time, Faraday uses an iron ring and wraps a wire connected to a battery around the left side of the ring. He wraps another wire around the right side of the ring and around a compass. When the current flows through the wire on the left, the iron ring becomes magnetized and he observes a deflection of the compass! This means that the magnetization of the iron ring has created an electric current in the second wire that creates, in turn, a magnetic field that interacts with the compass causing the deflection. He then turns his attention to electrolysis.

FIGURE 4.7   Michael Faraday (1791–1867).

In 1834, in his "Experimental Researches, 7th Series" he introduces several concepts and definitions that, for many of them, are still used nowadays in modern electrochemistry (see Figure 4.8)! Chemical compounds that are "decomposed directly by the electric current, their elements being set free" are called electrolytes (Greek: ἤλεκτρον (= electron) and λυτός (*lytos*)), meaning "which can be electrically decomposed". "Electrochemically decomposed" is renamed "electrolyzed" by Faraday. "The poles, as they are usually called, are only the doors of ways by which the electric current passes into and out of the decomposing body". He adds, "In place of the term pole, I propose using that of Electrodes, and I mean thereby that substance, or rather surface, whether of air, water, metal, or any other body, which bounds the extent of the decomposing matter in the direction of the electric current". The word "electrodes" comes from the Greek ἤλεκτρον (*electron*=electron) and ὁδός (*odos*=doorway). "Wishing for a natural standard of electric direction […] we propose calling that towards the east the anode, and that towards the west the cathode. The anode is therefore that surface at which the electric current according to our present expression, enters; it is the negative extremity of the decomposing body; is where oxygen, chlorine, acids, etc., are evolved; and is against or opposite the

On the Active Battery   227

in part, known and understood; but as their importance, and that of certain other coincident results, will be more evident by reference to the principles and experiments already stated and described, I have thought it would be useful, in this investigation of the voltaic pile, to notice them briefly here.

771. When the battery is in action, it causes such substances to be formed and arranged in contact with the plates as very much weaken its power, or even tend to produce a counter current. They are considered by Sir Humphry Davy as sufficient to account for the phenomena of Ritter's secondary piles, and also for the effects observed by M. A. de la Rive with interposed platina plates.[1]

772. I have already referred to this consequence (739) as capable, in some cases, of lowering the force of the current to one-eighth or one-tenth of what it was at the first moment, and have met with instances in which its interference was very great. In an experiment in which one voltaic pair and one interposed platina plate were used with dilute sulphuric acid in the cells, fig. 63, the wires of communication were so arranged that the end of that marked 3 could be placed at pleasure upon paper moistened in the solution of iodide

Fig. 63.

of potassium at x, or directly upon the platina plate there. If, after an interval during which the circuit had not been complete, the wire 3 were placed upon the paper, there was evidence of a current, decomposition ensued, and the galvanometer was affected. If the wire 3 were made to touch the metal of p, a comparatively strong sudden current was produced, affecting the galvanometer, but lasting only for a moment; the effect at the galvanometer ceased, and if the wire 3 were placed upon the paper at x, no signs of decomposition occurred. On raising the wire 3, and breaking the circuit altogether for a while, the apparatus resumed its first power, requiring, however, from five to ten minutes for this purpose; and then, as before, on making contact between 3 and p, there was again a momentary current, and immediately all the effects apparently ceased.

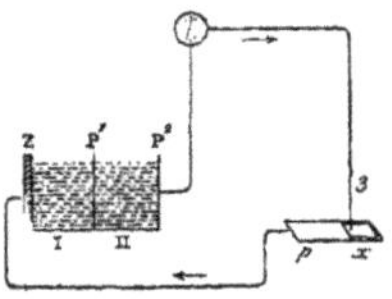

773. This effect I was ultimately able to refer to the state of the film of fluid in contact with the zinc plate in cell I. The acid of that film is instantly neutralised by the oxide formed;

[1] *Philosophical Transactions*, 1826, p. 413.

FIGURE 4.8   Faraday's active battery.

positive electrode. The cathode is that surface at which the current leaves the decomposing body, and is its positive extremity; the combustible bodies, metals, alkalis, and bases, are evolved there, and it is in contact with the negative electrode." Indeed, Faraday interprets the magnetism of the terrestrial globe as due to electric currents running in lines of latitude or, in other words, from east to west. In fact, "anode" comes from the Greek ἀνά (*ana* = upwards) and ὁδός (*odos* = doorway), meaning the way which the sun rises (east). Similarly, "cathode" is κατά (*kata* = downwards) and ὁδός (*odos* = doorway), the way which the sun sets (west). "If you take anode and cathode, I would propose for the two elements resulting from electrolysis the terms anion and cation […]; and for the two together you might use the term ions […]. The anion is that which goes to the anode, the cation is that which goes to the cathode." Cations come from the Greek (κατιόν = *kation*), meaning "that which goes down" and anions from (ἀνιόν = *anion*), "that which goes up". Ion from the Greek participle "which goes" thus denotes chemical that moves as a result of an electric current. Finally, in "Experimental Researches, 3rd series", Faraday posits that "when electrochemical decomposition takes place, there is great reason to believe that the quantity of matter decomposed is not proportional to the intensity, but to the quantity of electricity present". Let us add that, in the 10th series, Faraday introduces the "volta-electrometer". According to Faraday: "It offers the only actual measurer of voltaic electricity which we at present possess". Thanks to this framework, Faraday states his famous laws of electrolysis! The first law of electrolysis is as follows: "The chemical power of a current of electricity is in direct proportion to the absolute quantity of electricity which passes". In other words, the mass of a chemical decomposed at an electrode is directly proportional to the quantity of electricity used. The second law states: "Electrochemical equivalents coincide, and are the same, with ordinary chemical equivalents". It means that the mass of the substance decomposed at the electrode is directly proportional to the substance's equivalent weight. In practice, we use the Faraday to denote the quantity of electricity that causes a chemical change of one equivalent weight. One Faraday is equal to 9.6485×10^4 Coulombs of electricity. Thus, in the case of the electrolysis of calcium chloride ($CaCl_2$), 1 Faraday of electricity results in the deposition of 20.039 g (40.078/2) of calcium at the negative electrode and evolves 35.453 g of chlorine at the positive electrode. Let us add that Faraday goes on to make many more discoveries. Indeed, he invents the "Faraday cage" in 1836, introduces the so-called "Faraday effect" in 1845 and identifies the diamagnetism of all matter. Even living things, such as frogs, are diamagnetic and can be levitated in a strong magnetic field! In this case, the diamagnetic repulsion of the living tissue can balance exactly the effect of gravity throughout the body, resulting in a levitated frog!

CONDUCTIVITY OF SOLUTIONS AND DISSOCIATION THEORY

In 1853, Hittorf (1824–1914) publishes an article "On the Migration of Ions during Electrolysis". He builds on Faraday's ideas, according to whom electrolysis results from "the effects to arise from forces which are internal, relative to the matter under decomposition, and are not external, as they might be considered, if directly dependent on the poles", meaning that "the effects are due to a modification, by the electric current, of the chemical affinity of the particles through or by which that current is passing". Hittorf thus proposes that ions carry charges and migrate (or move) during electrolysis. His idea is that "the first action of the current consists in bringing the particles of the body to be decomposed into such a position that the cation of each molecule is turned towards the cathode, and the anion towards the anode. The two ions then separate from each other, move in opposite directions, and thereby meet with the neighboring ions likewise migrating." To better analyze the migration of ions, he introduces the concept of ion transport number and develops a method to measure it during electrolysis, now called Hittorf's method: "we shall find that the ions in each portion are in different proportion after electrolysis has taken place than before. This proportion is determined by the distance through which each ion moves during the passage of the current." During his experiments, he notes that some ions move faster than others! And he notes that the differences in concentrations at the electrodes are due to the differences in velocity between unlike species of ions. By means of a series of experiments, he succeeds in assigning transference or transport numbers, *Hittorfsche Überführungszahlen* in German, to many ions. For example, in the case of the electrolysis of copper sulfate (Experiment A in his article), he finds that 0.2955 g of Cu is deposited on an electrode after four hours of continuous current. Moreover, the solution around the cathode contains 2.2782 g of Cu before electrolysis and 2.0670 g of Cu after electrolysis. This means that the amount of Cu around the electrode decreases by 0.2112 g. The amount of transferred copper is thus 0.0843 g or, in other words, 28.5% equivalent, or transference for Cu ions for these experimental conditions. According to Hittorf, the sum of the transport numbers for all ions should be equal to one. This means that the sulfate ion has a transference of 71.5% equivalent. Differently put, the sulfate ions migrate faster than the copper ions! Also, Hittorf finds that there are different types of behaviors of ions in solutions; some ions (metal ions) can deposit on an electrode while others (sulfate) always remain in solution.

However, an astonishing idea is developed by Arrhenius (1859–1927) during his 1884 dissertation. What if those reactions were possible without any electric current? As we know from Faraday, chemical compounds can be "decomposed" and "set free" by an electric current, forming electrolytes. Arrhenius' idea is that chemical reactions in solutions are just reactions between ions and that the law of mass action applies, meaning that the rate of a chemical reaction is equal to the product of the masses of the reactants. According to him, solid salts can dissociate into pairs of charged ions in solution even in the absence of an electric current! For instance, when table salt (NaCl) is put into water, it dissolves, giving rise to Na^+ (positively charged sodium ions) and Cl^- (negatively charged chloride ions). The dissociation reaction is then written as $NaCl$ (s) $\rightarrow Na^+(aq) + Cl^-(aq)$. The presence of ions allows the solution to conduct electricity when subjected to an electric current, as in electrolysis.

Nowadays, electrolytes such as sodium, potassium, chloride, calcium and magnesium are used in energy drinks. Indeed, in physiology, Na^+, K^+, Ca^{2+}, Mg^{2+} and Cl^- are very important. For instance, sodium and chloride are found outside the cell whereas potassium, magnesium and calcium are found inside the cell. The subtle balance between these electrolytes allows for the regulation of blood pressure, body hydration, nerve and muscle functions among other things. For example, muscle contraction requires high enough levels of calcium, sodium and potassium, and, without them, muscle weakness and cramps may occur. The same is true with sodium and potassium during dehydration – this is the reason why energy drinks are usually full of sodium and potassium!

Despite the poor reception of his work, Arrhenius sends copies of his work to Ostwald and van't Hoff. Let us recall that van't Hoff (1852–1911) is the inaugural winner of the Nobel Prize in chemistry in 1901 "in recognition of the extraordinary services he has rendered by the discovery of the laws of chemical dynamics and osmotic pressure in solutions". As for Ostwald (1853–1932), he is the Nobel Prize awardee in 1909 in chemistry "in recognition of his work on catalysis, and for his investigations into the fundamental principles governing chemical equilibria and rates of reaction".

FIGURE 4.9   Jacobus Henricus van't Hoff (1852–1911).

Indeed, van't Hoff (see Figure 4.9) publishes in 1884 a book that revolutionizes the study of chemical reactions: *"Etudes de dynamique chimique"* (Studies in Chemical Dynamics). He proposes an entirely new framework to understand chemical reactions. He makes a distinction *"entre la transformation chimique totale et la transformation chimique limitée"* (between a complete chemical reaction and an incomplete chemical reaction or chemical equilibrium). He also defines a complete reaction as *"La premiere s'énonce en peu de mots; c'est la transformation chimique généralement connue, caracterisée par un changemenent total d'une substance (système initial) en d'autres qui en diffèrent (système final)"*. (The first can be described in very few words: it is the chemical transformation as commonly known, characterized by a complete change of a substance (initial system) into others of different nature (final system).) For chemical equilibrium, he notes that *"la transformation limitée [...] est caracterisée parce qu'elle s'arrête avant d'être totale. Dans l'état final, on trouve donc une partie des corps primitifs inaltérés en présence de corps nouvellement formés"*. (The incomplete transformation is characterized by the fact that it stops before completion. In the final state, one thus finds parts of the initial substances together with newly formed substances.) Also, he generalizes the concept of dynamical equilibrium to chemical reactions. This means that a chemical reaction can be actually seen as the combination of two equations, one going forward $(A+B \rightarrow C+D)$ and the other going backward $(A+B \leftarrow C+D)$, the equilibrium being written as

$A + B = C + D$. He also uses the law of mass action to determine the rate of a chemical reaction, v defined as $v = k \cdot [A]^{\alpha} [B]^{\beta}$ in which k is the rate constant, $[A] = m_A/V$, $[B] = m_B/V$, α and β are exponents. In addition, he states that *"nous allons nous servir, en un mot, du lien que forme cet équilibre entre l'étude de la marche de la transformation et la thermodynamique"*. (We are going to take advantage, in short, of the connection made by this equilibrium between the study of the rate of the transformation and thermodynamics.) This allows him to connect the amounts of reactants (A and B in the example above) and the products (C and D) to the ratio of the rate constant in the forward direction over the rate constant in the backward direction. He also notices that *"l'influence de la température sur la vitesse de la transformation varie avec la réaction que l'on considère"*, meaning that when the temperature increases, the rate of the reaction increases too! He also theorizes the concept of chemical affinity which is *"la force qui produit une transformation chimique"* (the force that produces a chemical transformation). When a chemical system reaches equilibrium, he finds that the overall activity, or in other words, the difference between the affinities of the products and reactants is equal to zero! Furthermore, he observes that *"en passant par le point de transition, cette affinité ou cette différence d'affinité change de signe"*. (When passing by the transition point, the overall affinity, or affinity difference, reverses its sign.) In other words, the sign of the affinity tells in which direction (forward or backward) it evolves to reach equilibrium. This is an amazing finding since it allows the prediction of what reaction is going to occur!

As for Ostwald (see Figure 4.10), he also studies reaction rates and defines a new phenomenon known as catalysis. According to him, "catalysis is the acceleration of a slowly proceeding chemical reaction through the presence of a foreign substance". He gives this foreign substance the name "catalyst" and defines the two fundamental principles of catalysis. First, he observes that "the acceleration occurs without a change of the energetic situation". In other words, the catalyst does not change

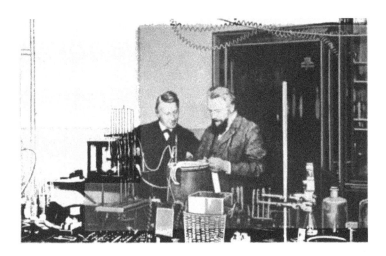

FIGURE 4.10 Wilhelm Oswald and Jacobus Henricus van't Hoff.

the affinity of the reactants and products and thus, does not impact the amount of product formed: it only makes the chemical reaction occur faster. Second, "at the end of the chemical reaction the foreign substance can be considered as removed". This means that the products are the same as in the absence of the catalyst, and that the catalyst is entirely released at the end of the reaction. No catalyst is thus used up and very small amounts of a catalyst are therefore required to accelerate a chemical reaction! This revolutionizes the chemical industry! For instance, Ostwald develops catalysts to produce fertilizers through the oxidation of ammonia to nitric acid – a process that is still used nowadays. This also opens the door to a new understanding of the metabolism of living systems. In biology, catalysts are often called enzymes, which enable the breaking down of large molecules into smaller pieces that can be easily absorbed by the body. This is the case, for instance, of lactase, which speeds up the splitting of a large sugar molecule, lactose, into two smaller molecules, galactose and glucose, that can be readily metabolized by the human body.

In 1887, Ostwald and van't Hoff establish the *"Zeitschrift für Physikalische Chemie"* (ZPC). It is an international journal of research in physical chemistry and chemical physics that is still being published. Finally, let us add the famous "greenhouse law" stated by Arrhenius in 1896: "If the quantity of carbonic acid increases in geometric progression, the augmentation of temperature will increase nearly arithmetic progression". In other words, according to Arrhenius, if the quantity of CO_2 in the atmosphere, for example, is multiplied by 2, then the temperature will rise by 5°C! He receives the Nobel Prize in Chemistry in 1903 "in recognition of the extraordinary services he has rendered to the advancement of chemistry by his electrolytic theory of dissociation".

ELECTROLYSIS AND THE "ATOM" OF ELECTRICITY

Stoney (1826–1911) introduces the concept of a unit of electric charge in 1874. He identifies that there is a quantity of electricity involved in chemical bonds, such as between H and Cl in HCl, or between H and O in H_2O. According to him, during the electrolysis of solutions, this quantity of electricity then passes when bonds are broken. In particular, he states "Nature presents us, in the phenomenon of electrolysis, with a single definite quantity of electricity which is independent of the particular bodies acted on. [...] For each chemical bond which is ruptured within an electrolyte a certain quantity of electricity traverses the electrolyte which is the same in all cases. This definite quantity of electricity I shall call E_1. If we make this our unit quantity of electricity, we shall probably have made a very important step in our study of molecular phenomena..." In 1894, he adds, "In this paper an estimate was made of the actual amount of this most remarkable fundamental unit of electricity, for which I have since ventured to suggest the name electron. According to this determination the electron = a twentiethet (that is 10^{-20}) of the quantity of electricity, which was at that time called the Ampere, viz.: the quantity of electricity which passes each second in a current of one Ampere." Indeed, he finds that 1 Ampere-second (or Coulomb) results in the dissociation of 92×10^{-6} g of H_2O, which gives him 1.15×10^{20} H atoms and thus a unit of electricity of 1×10^{-20}!

For his part, von Helmholtz (1821–1894) thinks that electricity is composed of "atoms of electricity". Indeed, in 1881, he writes: "If we accept the hypothesis that the elementary substances are composed of atoms, we cannot avoid concluding that electricity also, positive as well as negative, is divided into definite elementary portions which behave like atoms of electricity. As long as it moves about on the electrolytic liquid each ion remains united with its electric equivalent or equivalents. At the surface of the electrodes decomposition can take place if there is sufficient electromotive force, and then the ions give off their electric charges and become electrically neutral."

Another breakthrough is made by Nernst (1864–1941). He is Ostwald's chief assistant at Leipzig and is particularly interested in galvanic cells and in their connections with thermodynamics. He is especially keen to understand where the "seat of the electromotive force" is (what makes the ions migrate). He then builds on von Helmholtz's work which establishes that putting in motion the charges requires work and this work is called "electromotive force (emf)". In 1889, Nernst demonstrates that everything happens at the interfaces between the electrodes and the electrolyte! He develops the famous "Nernst equation" that links thermodynamics and the chemical reactions occurring in a battery. First, he starts by calculating the work produced by the chemical reactions and writes $W_{reaction} = RT\ln Q - RT\ln K$ in which $Q = [products]_t/[reactants]_t$ is the reactional quotient for a specific time t during the reaction. $K = [products]_{eq}/[reactants]_{eq}$ is the equilibrium constant. The presence of an electric current implies that there is an electrical work that results from the chemical reactions, meaning that the $W_{reaction}$ can be found through the electrical work! In other words, $W_{electrical} = W_{reaction}$. Since $W_{electrical} = -nFE$ (E is the electromotive force, F the Faraday constant and n is the number of electrons exchanged at the interface) and $W_{reaction} = RT\ln Q - RT\ln K$, we end up with the following equation: $-nFE = RT\ln Q - RT\ln K$. Dividing both sides by $-nF$, it follows that $E = (RT/nF) \ln(K/Q)$. Let us note that at equilibrium $K = Q$, which means that $E_{eq} = 0$. In practice, equilibrium is reached after a long time, which is what we see with old batteries: they do not work anymore! Finally, Nernst receives the 1920 Nobel Prize in chemistry "in recognition of his work in thermochemistry".

However, one of the voltaic battery's issues is that hydrogen bubbles form on the electrode, limiting its active surface area. This is the reason why Daniell (1790–1845) invents a new battery in 1836, the well-known "Daniell cell" (see Figure 4.11) which solves this problem. Daniell receives the Copley Medal in 1837 in recognition for this invention! In practice, a Daniell battery consists of a container divided into two compartments, connected via a membrane permeable to ions (or bridge salt which allows for the passage of ions from one compartment to another). The first compartment is filled with zinc sulfate ($ZnSO_4$) solution, in which a zinc (Zn) electrode is dipped. The second compartment is filled with a copper sulfate ($CuSO_4$) solution, in which a copper (Cu) electrode is dipped. In electrochemistry, a short-hand notation is Zn(s) | $ZnSO_4$ (aq) || $CuSO_4$(aq) | Cu(s). As in Volta's battery, the Zn anode undergoes a decomposition and gives rise to zinc ions (Zn^{2+}). The chemical equation reads $Zn \rightarrow Zn^{2+} + 2e^-$. As for the second compartment, the copper ions (Cu^{2+}) deposit on the Cu cathode according to $Cu^{2+} + 2e^- \rightarrow Cu$. The total chemical equation for the container (the two compartments) is therefore given by $Zn + Cu^{2+} + 2e^- \rightarrow Zn^{2+} + 2e^- + Cu$.

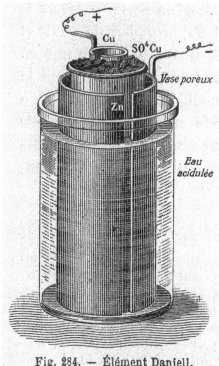

Fig. 284. — Élément Daniell.

FIGURE 4.11 Daniell cell.

The $2e^-$ appear on both side and thus cancel out giving $Zn + Cu^{2+} \rightarrow Zn^{2+} + Cu$. From a thermodynamic standpoint, the equilibrium constant for this reaction is $K = [Zn^{2+}]_{eq}/[Cu^{2+}]_{eq}$. The reactional quotient Q is $Q = [Zn^{2+}]_t/[Cu^{2+}]_t$. This means the emf E for the Daniell cell depends on the actual concentration in Zn^{2+} ions and Cu^{2+} ions in the two compartments! For instance, if the two initial (t = 0) concentrations are equal then $Q = 1$ and knowing that $K = 1.64 \times 10^{37}$, we therefore have $E = (RT/nF) \ln(K/1) = 1.1$ V. Now if we have a hundred times more Zn^{2+} ions than Cu^{2+} ions, we then have $Q = 100$, meaning that this time, the battery produces 0.4 V! The Daniell cell becomes very popular and serves as a source of electricity for the telegraph!

During the 20th century, a new source of clean energy emerges based on electrochemical processes. It is called fuel cells and works like a battery (see Figure 4.12). It relies on the supply of hydrogen to the anode and of air to the cathode. The full process only emits water! Currently, the most widely used type of fuel cell is the proton-exchange membrane fuel cell (PEMFC). In PEMFC, hydrogen reacts at the anode according to the chemical reaction: $2 H_2 \rightarrow 4 H^+ + 4 e^-$ while the oxygen from the air reacts at the cathode according to $O_2 + 4 H^+ + 4 e^- \rightarrow 2 H_2O$. This means that the overall reaction, when we add up the two reactions occurring at the cathode and anode, can be written as $2 H_2 + O_2 \rightarrow 2 H_2O$. Only water is indeed produced! The PEMFC takes its name from the electrolyte that is used in the cell: it is a special material called a proton exchange membrane that only allows positively charged

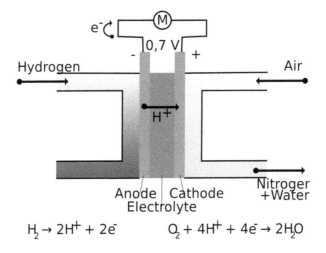

$$H_2 \rightarrow 2H^+ + 2e^- \qquad O_2 + 4H^+ + 4e^- \rightarrow 2H_2O$$

FIGURE 4.12 An example of a fuel cell.

ions to migrate. Furthermore, platinum nanoparticles are used both at the cathode and at the anode as catalysts to accelerate the chemical reactions occurring at the electrodes. The efficiency of fuel cells is very high, typically around 50%, compared to roughly 25% for traditional internal combustion engines. PEMFC are thus very efficient and can be used in mobile applications to power vehicles. For instance, fuel cell buses are running in California, British Columbia and in large European cities like Barcelona, London, Madrid, Stockholm and Stuttgart!

ORDER AND PERIODICITY IN THE PROPERTIES OF ELEMENTS

For centuries, different techniques, such as combustion, distillation and electrolysis, are developed to better understand what constitutes Nature and, more precisely, the substances and compounds that surround us. The classification of elements that form chemical substances or compounds develops then very rapidly as chemistry evolves. An early classification of "simple substances" is proposed by Lavoisier in 1789. Indeed, in his *"Traité élémentaire de Chimie"*, he writes on his *"Observations Sur le Tableau des Substances fimples, ou du moins de celles que l'état actuel de nos connoiffances nous obligent à confidérer comme telles"*. He comes up with a table that groups these simple substances according to their properties into gases, non-metals, metals and earths. Lavoisier is also aware that this is just a first attempt at a classification: *"La Chimie marche donc vers fon but & vers fa perfection, en divifant, fubdivifant, & refubdivifant encore, & nous ignorons quel fera le terme de fes fucces. Nous ne pouvons donc pas affurer que ce que nous regardons comme fimple aujourd'hui le foit en effet."* (Chemistry progresses towards its goal and towards its perfection, in dividing, subdividing, and resubdividing again, and we don't know yet what the final success will be. We therefore cannot assert that what we are considering as simple today is really simple indeed.)

About 30 years later, Döbereiner (1780–1849) introduces the concept of triads of elements that have similar chemical properties. For instance, he groups together lithium (Li), sodium (Na) and potassium (K) by noticing that all three react with water at room temperature and all form compounds with chlorine (Cl) with similar formulae (LiCl, NaCl, KCl)! Furthermore, he finds that the chemical properties, such as, for instance, the weight, of the middle element (Na) can be found by averaging those of the other two (Li and K). He also discovers four other triads! One of them is calcium (Ca), strontium (Sr) and barium (Ba), with the middle element being Sr.

In 1862, the French geologist de Chancourtois (1820–1886) is the first to arrange atoms periodically in order of their atomic weights, writing them on the outside of a cylinder (see Figure 4.13). He also decides that oxygen will be the standard for his apparatus, meaning that when a complete turn of the cylinder is done, it corresponds to an atomic weight increase of 16. When rotating the cylinder, he observes that all

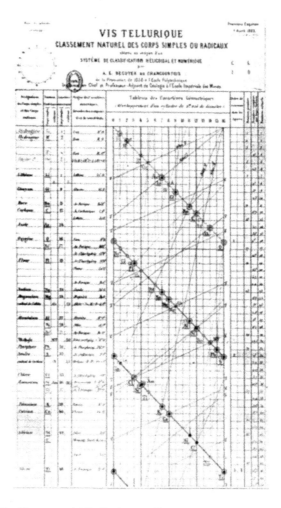

FIGURE 4.13   De Chancourtois' "telluric screw".

elements fall onto a spiral! And he sees that elements with similar properties can be found by plotting a vertical line on the cylinder! He calls his apparatus the "Telluric screw" which represents the first apparatus to classify the elements!

The next step is made by Newlands (1837–1898). He follows up on de Chancourtois' work and orders the elements according to their atomic weights. He numbers the elements as 1, 2, 3 and so on and states that "it will be observed that elements having consecutive numbers frequently either belong to the same group or occupy similar positions in different groups". Newlands starts to identify a periodicity in his list of elements: "The difference between the number of the lowest member of a group and that immediately above it is 7; in other words, the eighth element starting from a given one is a kind of repetition of the first, like the eighth note of an octave in music. The differences between the numbers of the other members of a group are frequently twice as great; thus in the nitrogen group, between N and P there are seven elements; between P and As 13; between As and Sb, 14 and between Sb and Bi, 14." He proposes the "law of octaves" to account for this periodicity, in analogy with the octaves of music!

Next, in 1868, Meyer (1830–1895) proposes a new table in which atoms are arranged in order of their weights. He also comes up with a graph that shows the atomic volume as a function of the weight. This is quantitative proof that the properties of atoms depend periodically on their weights. Meyer's table is very similar to the one we now know as Mendeleev's periodic table. However, because of a delay in publication, Mendeleev's classification appears first!

Mendeleev (1834–1907) receives the prestigious Copley Medal in 1905 "for his contributions to chemical and physical sciences". To compose his periodic table (see Figure 4.14), he starts by writing down the properties of elements on cards. He then

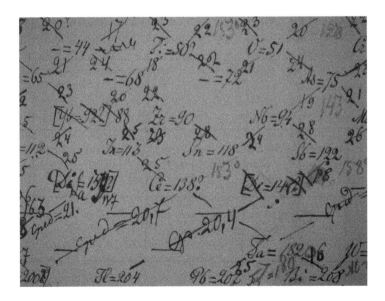

FIGURE 4.14   Mendeleev's classification.

orders them according to their atomic weights and rearranges them in vertical columns, in which elements have similar chemical properties. One of Mendeleev's intuitions is that the elements should not be arranged strictly in ascending order of their weights – chemistry should trump atomic weights! For instance, he finds for iodine (I) an atomic weight that is larger than for tellurium (Te). However, in the periodic table, he puts tellurium in the same column as oxygen (O), sulfur (S) and selenium (Se), exactly one column before iodine (I). Indeed, iodine has similar properties to fluorine (F), chlorine (Cl) and bromine (Br) which constitute what we now call the halogen group. Another genius idea of Mendeleev is to leave empty slots for elements that have not yet been discovered. However, while living the name blank, he predicts the properties these unknown elements have. Remarkably, a few years later, he is proved right, first with the discovery of gallium (Ga) in 1875, then with the discovery of scandium (Sc) in 1879 and the discovery of germanium (Ge) in 1885.

Yet, over the years, several scientists make new discoveries that initially appear to challenge Mendeleev's table. For instance, Ramsay and Lord Rayleigh discover a new element called argon (Ar) in 1894. They state that "If argon be a single element then there is reason to doubt whether the periodic classification of elements is complete; whether in fact elements may not exist that cannot be fitted among those of which it is composed". In the following years, Ramsay finds four other elements: helium (He), neon (Ne), krypton (Kr) and xenon (Xe). These elements are now called noble gases. They are very different from other elements since they do not seem to react with the other elements. In addition, they do not seem to fit within the period of seven elements from the first line of the periodic chart to the second one. A few years later, the solution is found, and a new column is added to the periodic chart between the halogens and the alkali metals (Li, Na, K, etc.) groups. Mendeleev is elated: "This was extremely important for [Ramsay] as an affirmation of the position of the newly discovered elements, and for me as a glorious confirmation of the general applicability of the periodic law". A second challenge appears with the proposal by van den Broek that elements should be ordered according to their nuclear charge instead of their atomic weight.

A physicist from the University of Manchester, Moseley (1887–1915) works for Rutherford and carries out a series of experiments to test van den Broek's idea. Moseley analyzes the X-ray spectrum given off by elements on the same line in the periodic table. He also focuses on the places for which "chemical order" inverts the order of atomic weights such as, for instance, K-Ar, Ni-Co and I-Te. Specifically, he looks at the frequencies for which K lines (see Fraunhofer) are observed. He finds that theses frequencies are proportional to the squares of the numbers corresponding to the position of the element in the periodic chart! He concludes that "we have here a proof that there is in the atom a fundamental quantity, which increases by regular steps as we pass from one element to the next. This quantity can only be the charge of the central positive nucleus, of the existence which we already have definite proof." With the discovery of the proton in 1920, this quantity is identified as the number of protons in the nucleus and called Z. As mentioned before, it also corresponds to the position number of an element in the periodic chart (German: *zahl* = number). Interestingly, it is also equal to the number of electrons in the atom since an atom should be neutral!

CHEMICAL BONDING AND LEWIS STRUCTURES

At the beginning of the 20th century, the electron has been observed and its properties have been measured (mass, charge, etc.). Many chemists are keen to apply the electrochemical theory to explain all "chemical unions". Indeed, they think that the electrolytic dissociation of sodium chloride (NaCl) into two parts, one negative (Cl⁻) and one positive (Na⁺), can be the beginning of an explanation. In such substances, they assume that, at all times, there is a significant displacement of electricity from sodium to chlorine, leading to the dissociation into sodium and chlorine ions. This displacement of electricity is called "polarity" and substances like NaCl are termed as "polar" substances. However, some substances like, for instance, the molecule of hydrogen (H_2) or methane (CH_4) do not show such an electricity displacement and for this reason are called "non-polar". This brings Lewis (1875–1946) to ask the following questions: "Must we conclude that there are two distinct types of chemical union, one a completely polar and the other a completely non-polar type? [...] Or can we find some means of ascribing all the most varied types of chemical union to one and the same fundamental cause, differing only in the nature and degree of its manifestation?" To answer this question, he proposes a "modern dualistic theory" in 1913. He starts by introducing the concept of the "valence" of an atom as "the valence of an atom in an organic molecule represents the number of bonds which tie this atom to other atoms. Moreover in the mind of the organic chemist, the chemical bond is no mere abstraction; it is a definite physical reality, a something which binds atom to atom. Although the nature of such a tie remained mysterious, yet the hypothesis of the bond was amply justified by the signal adequacy of the simple theory of molecular structure to which it gave rise." Unlike Berzelius, who thinks that electricity is a fluid that flows from an atom to another, Lewis proposes a new picture: "Thus the modern dualistic theory regards chemical action as primarily due to the jumping of electrons from atom to atom". For example, electrolytes demonstrate this concept of separation of charges. In aqueous solutions of NaCl, the sodium atom (Na) loses an electron to become the sodium cation (Na⁺) while the chlorine atom (Cl) gains an electron and becomes the chloride anion (Cl⁻). Thus, Lewis goes on to posit that a crystal of NaCl is already composed of Na⁺ and Cl⁻ ions which are held together by electric forces. NaCl is thus called an ionic solid. However, this theory of valence does not explain everything. Indeed, it does not work for the relatively non-polar compounds found in organic chemistry, since these compounds are not an "assemblage of charged atoms". He concludes, "If the properties of substances could not be explained by the mere assumption of charged atoms, might they not be explicable if we should no longer regard the atom as a unit, but rather if we might ascertain where the charge or charges resided within the atom itself?" In 1907, Thomson starts to draw a picture of chemical bonding that accounts for what happens, for instance, in the H_2 molecule. He considers the case of two spheres equal in size and overlapping with each other, with electrons located symmetrically in the region of overlapping: "In this case there is no difference in the electrification of the spheres; we cannot say that one is positively, the other negatively electrified; and if the spheres were separated after having been together they would each be neutral [...] We thus see that it is possible to have forces electrical in their origin binding the two systems together without a resulting charge on either system." Several scientists including Stark, Kossel and Parson reach the same conclusion: "two atoms are held together by groups of electrons held in common by the two atoms"! And some of them even find that "the group of bonding electrons is a group of two".

In his paper "The Atom and the Molecule" published in 1916, Lewis introduces his new theory of valence to explain the chemical bond. In particular, he notices with Parson that atoms, when they gain or lose electrons to become ions, often acquire an arrangement of eight outer electrons. They call it the "group of eight", renamed a few years later as "octet" by Langmuir. Lewis' new theory can be summed up as "Two atoms may conform to the rule of eight, or the octet rule, not only by the transfer of electrons from one atom to another, but also by sharing one or more pairs of electrons. These electrons which are held in common by two atoms may be considered to belong to the outer shells of both atoms." To better understand his theory, Lewis introduces a new notation in which he represents each valence electron by a dot. One can see that there are eight dots around each of the "heavier" atoms, C and Cl, which is a manifestation of the rule of eight or octet rule. He also cautions that this two-dimensional representation does not account for the stereochemistry of molecules. Most of Lewis' structures show that the regular state of a molecule involves each atom of hydrogen surrounded by a pair of electrons while every other atom has a group of eight electrons around it. This is achieved by sharing a pair of electrons or bond between two atoms. However, there are many cases in which having only a single pair of electrons between atoms does not result in structures following the octet rule. It is at this point that Lewis envisions the concept of double and triple bonds. An example for CO_2 (carbon dioxide) is presented in Figure 4.15. Indeed, in the case of the O_2 molecule, each atom of O has six valence electrons and "even if two atoms share an electron pair this does not suffice for the completion of two octets". Lewis' solution is the following: "if two pairs of electrons are shared, then each atom can be said to have its group of eight". Thus, the double bond actually stands for two pairs of electrons shared between two atoms. Then, for the C_2H_2 molecule, Lewis draws a triple bond between the two carbon atoms and states "We should represent a triple bond by three pairs of electrons held in common by two atoms. This is the only way in which we can assign to each carbon in acetylene its full quota of eight electrons". Lewis' theory for chemical bonding is very general and not only applies to molecules but also to ions. For instance, adding a hydrogen ion (H^+) to the nitrogen atom in ammonia (NH_3) results in an ammonium ion (NH_4^+). The same is true in adding a hydrogen ion (H^+) to water (H_2O) which gives a hydronium ion (H_3O^+). Finally, Lewis extends the concept of electron sharing and valence to inorganic compounds

FIGURE 4.15   Lewis structure for the CO_2 molecule: O=C=O.

as well and states "therefore we may define the valence of an atom in any molecule as the number of electron pairs which it shares with other atoms". At the same time, Langmuir (1881–1957) works on chemical bonding and calls this number of bonds "covalence". This is the reason why shared pairs of electrons are now called covalent bonds!

As mentioned above, the Lewis structure only gives a picture in 2D of a molecule. The 3D organization of a molecule can be deduced using the valence shell electron pair repulsion (VSEPR) method introduced by Gillespie and Nyholm in 1957. Starting from an observation made by Lewis on methane, they realize that four electron pairs around the carbon atom give rise to a tetrahedral geometry for the molecule of methane. Also, in the 1940s, work by Sidgwick and Powell shows that both bonding and nonbonding pairs of electrons could play a role in the 3D geometry of a molecule. Gillespie and Nyholm then propose a model in which molecules are classified into two types of molecules: AX_n and AX_nE_m where A is the central atom, X is another atom bonded to A and E is a lone pair (n and m are two integers representing the numbers of X and E, respectively). The idea behind VSEPR is that all electron pairs around A, or in other words, in A's valence shell, repel each other and are arranged as far as possible from one another. For instance, the CO_2 molecule is of the type AX_2 and has a linear geometry. The H_2O molecule is of the type AX_2E which represents a bent geometry. More examples are given in Figure 4.16. Even if several exceptions have been uncovered over the years, VSEPR remains an

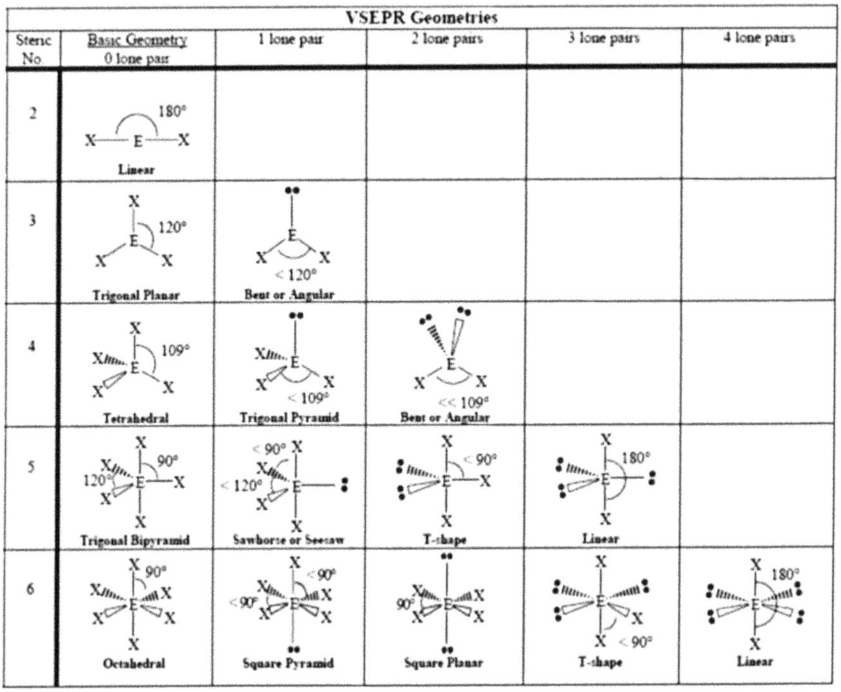

FIGURE 4.16   VSEPR theory for the determination of the 3D geometry of a molecule.

extremely useful tool to quickly predict the geometry of a molecule without using any quantum chemistry calculations.

VALENCE, OXIDATION NUMBER AND REDOX PROCESSES

After defining chemical bonding, chemists are now eager to apply the concept of valence to chemical reactions. They start by studying oxidation reactions that will be known later as reduction–oxidation or redox reactions. Hildebrand (1881–1983), who is also a professor at Berkeley, explains the connection between valence and oxidation. He writes in 1918: "The term oxidation is applied whenever valence takes on a more positive (or less negative) value. The opposite process [...] the decrease in valence, is called by the more obvious general name of reduction." This resonates with what Fry concludes in 1915: "The development of a positive valency by an atom (schematically through the loss of an electron) corresponds to oxidation. When an atom develops a negative valency (schematically through the gain of an electron) it is reduced." In 1938, Latimer (1893–1955) proposes a general framework in which he defines the concepts of "oxidation number" or "oxidation state", as well as "oxidation potential" in his book "The Oxidation States of the Elements and Their Potentials in Aqueous Solution". He classifies elements by their oxidation numbers (ON). For example, for pure elements such as Ag: ON $= 0$. For magnesium ion (Mg^{2+}), the oxidation number is ON $= +II$ (this means that it needs two electrons to form chemical bonds), and chloride ion (Cl^-) has an oxidation number ON $= -I$ (this means that it can give off one electron to form a chemical bond). By adding $2Cl^-$ to $1Mg^{2+}$, we can see that we have a total ON $= 0$, meaning that $MgCl_2$ exists on its own and is neutral! In other words, the oxidation state represents the number of electrons that an element can gain or lose when forming chemical bonds. According to International Union of Pure and Applied Chemistry (IUPAC) in 2016, the official definition of the oxidation state (OS) of an atom is "The Oxidation State of an atom is the charge of this atom after ionic approximation of its heteronuclear bonds". Furthermore, "An alternative term 'oxidation number' is also used in English. It is largely synonymous with OS and may be preferred when the value represents a mere parameter or number, rather than being related to chemical systematics or a state of the atom in a compound."

Redox chemical reactions are defined as all reactions in which atoms undergo a change in their oxidation numbers (or states). They are composed of two half-reactions, an oxidation process accompanied by a parallel reduction process. Indeed, an oxidation reaction is defined as a loss of electrons whereas reduction is a gain of electrons. In other words, an atom is oxidized when it loses electrons, while an atom is reduced when it acquires electrons.

One class of redox reactions relies on electron(s) transfer. Let us go back to the simple example of the Daniell cell. The anode is a zinc electrode (Zn) which undergoes the following oxidation half-reaction: $Zn \rightarrow Zn^{2+} + 2e^-$. In this equation, the reactant Zn has an oxidation number ON(Zn) $= 0$, while the product Zn^{2+} has an oxidation number ON(Zn^{2+}) $= +II$. Here we deal with an oxidation reaction, since we have in increase in ON ($0 \rightarrow +II$) associated with the release of the two electrons. On the other hand, at the cathode, there is a deposit of copper (Cu) that forms according to this reduction half-reduction: $Cu^{2+} + 2e^- \rightarrow Cu$. In this case, the reactant

Cu^{2+} has an $ON(Cu^{2+}) = +II$ and the product Cu has an $ON(Cu) = 0$. There is thus a decrease in ON ($+II \rightarrow 0$) meaning that we have a reduction reaction and that Cu^{2+} has gained two electrons in the process. Adding up the two half-equations gives the redox reaction: $Cu^{2+} + Zn \rightarrow Zn^{2+} + Cu$. As we can see, we have two redox couples: Cu^{2+}/Cu and Zn^{2+}/Zn. Using Latimer's tables, there is a standard electrochemical potential E_0 assigned to each redox couple. In this case, $E_0(Cu^{2+}/Cu) = 0.34$ V and $E_0(Zn^{2+}/Zn) = -0.76$ V. Let us add the difference between the two standard potentials, which gives 1.1 V, exactly the emf of the Daniell cell! The E_0 values also predict how the redox reaction takes place: the oxidizing agent (Cu^{2+}) of the redox couple with the highest E_0 reacts with the reducing agent (Zn) of the redox couple with the lowest E_0. This is called the "gamma rule". However, these reactions can also be very tricky. Let us look, for instance, at the redox reaction MnO_4^- and I^- which gives I_2 and Mn^{2+}. Starting with the redox couple MnO_4^-/Mn^{2+}, we know that $E_0(MnO_4^-/Mn^{2+}) = 1.49$ V and for the redox couple I_2/I^-, we know that $E_0(I_2/I^-) = 0.54$ V. Moreover, for the redox couple MnO_4^-/Mn^{2+}, $ON(MnO_4^-) = ON(Mn \text{ in } MnO_4^-) + 4 \, ON(O \text{ in } MnO_4^-) = -I$ and $ON(Mn^{2+}) = +II$. In general, ON (O in a compound) $= -II$; it then follows that $ON(Mn \text{ in } MnO_4^-) = +VII$. Equivalently, we have $VII + 4(-II) = -I$. Therefore, the oxidizing agent MnO_4^- in which $ON(Mn \text{ in } MnO_4^-) = +VII$ is reduced into $ON(Mn^{2+}) = +II$. Or, differently put, $VII \rightarrow II$, gaining five electrons. The reduction half-reaction is thus $MnO_4^- + 5e^- \rightarrow Mn^{2+}$, but the equation is not balanced. First, we need to add four oxygen on the right-hand side, which translates into the addition of four H_2O molecules: $MnO_4^- + 5e^- \rightarrow Mn^{2+} + 4H_2O$. Second, we need to balance the H atoms, which means adding $8H^+$ on the left-hand side. Finally, we see that the number of Mn, O and H atoms are balanced as well as the total charges on both sides of the equation. In conclusion, we have the following half-reaction: $MnO_4^- + 5e^- + 8H^+ \rightarrow Mn^{2+} + 4H_2O$. For the redox couple, I_2/I^- we have $ON(I_2) = 0$ and $ON(I^-) = -I$. The reducing agent $ON(I^-) = -I$ is oxidized into $I_2 \, ON(I_2) = 0$. In other words, $-I \rightarrow 0$, losing one electron. The oxidation half-reaction is then $I^- \rightarrow I_2 + 1e^-$, but the iodine atoms are not balanced. By adding one I^- on the left-hand side, we have $2I^- \rightarrow I_2 + 1e^-$. This time, the charges are not balanced anymore, which prompts the addition of one electron on the right-hand side. This leads to the other half-reaction: $2I^- \rightarrow I_2 + 2e^-$. To find the complete redox reaction, we add both half-reactions, giving: $MnO_4^- + 5e^- + 8H^+ + 2I^- \rightarrow Mn^{2+} + 4H_2O + I_2 + 2e^-$. Since this is a redox reaction, the number of electrons on both sides should be equal, due to electron transfer. This is the reason why the redox equation is written as $2MnO_4^- + 10e^- + 16H^+ + 10I^- \rightarrow 2Mn^{2+} + 8H_2O + 5I_2 + 10e^-$ or if we simplify: $2MnO_4^- + 16H^+ + 10I^- \rightarrow 2Mn^{2+} + 8H_2O + 5I_2$. The trick here is that, to obtain the same number of electrons on both sides, the reduction half-reaction should be multiplied by 2 and the oxidation half-reaction by 5 before summing them up.

Redox reactions are also very important in biology. For example, photosynthesis and cell respiration are two processes that rely on redox reactions. In the case of cell respiration, the first redox couple involved is $\frac{1}{2}O_2/H_2O$ with $E_0(\frac{1}{2}O_2/H_2O) = +0.82$ V and the second redox couple involved is $NAD^+/NADH$ with $E_0(NAD^+/NADH) = -0.32$ V. NADH stands for the reduced form of nicotinamide adenine dinucleotide (NAD). For the couple O_2/H_2O, we have $ON(O_2) = 0$ and $ON(O \text{ in } H_2O) = -II$. Therefore the oxidizing agent O_2 in which $ON(O_2) = 0$ is reduced into $ON(O \text{ in } H_2O) = -II$ or in other words, $0 \rightarrow -II$, meaning that we have a gain of $2e^-$. This gives a reduction half-reaction of $\frac{1}{2}O_2 + 2e^- \rightarrow H_2O$, but the number of hydrogen atoms is not balanced. Thus, we add $2H^+$ on the left-hand side. This leads to $\frac{1}{2}O_2 + 2e^- + 2H^+ \rightarrow H_2O$. For the redox couple $NAD^+/NADH$, we know that the difference $ON(NAD^+) - ON(NADH) = II$, meaning that the reducing agent NADH is oxidized in NAD^+ and releases $2e^-$. The oxidation half-reaction is $NADH + H^+ \rightarrow NAD^+ + 2H^+ + 2e^-$. When we add the two equations, we obtain the complete redox equation as $\frac{1}{2}O_2 + 2e^- + 2H^+ + NADH + H^+ \rightarrow NAD^+ + 2H^+ + 2e^- + H_2O$. Simplifying the equation, we finally obtain $\frac{1}{2}O_2 + NADH + H^+ \rightarrow NAD^+ + H_2O$. The NAD^+ is then used by the cell in a metabolic pathway, known as glycolysis or "sugar splitting", that produces energy in the form of two ATP molecules. In 1997, Boyer and Walker receive the Nobel Prize "for their elucidation of the enzymatic mechanism underlying the synthesis of adenosine triphosphate (ATP)". In the press release by the Nobel committee, "ATP functions as a carrier of energy in all living organisms from bacteria and fungi to plants and animals including humans. ATP captures the chemical energy released by the combustion of nutrients and transfers it to reactions that require energy, e.g. the building up of cell components, muscle contraction, transmission of nerve messages and many other functions. ATP has been termed the cell's energy currency."

ELECTRONEGATIVITY AND QUANTUM ELECTROCHEMISTRY

From 1925 on, the advent of quantum mechanics gives a new perspective on molecular structure and chemical bonding. Indeed, it is now possible to understand atoms and their union in molecules using only equations like the Schrödinger equation. In a series of papers beginning in 1931, Pauling (1901–1994) sheds light on the nature of the chemical bond. He also publishes a book in 1939, "The Nature of the Chemical Bond and the Structure of Molecules and Crystals: An Introduction to Modern Structural Chemistry", which is one of the most important books in the 20th century and a reference for a new generation of chemists around the world! He first investigates the electron-pair bonds and formulates a set of rules which links the strength of the bond and the single-electron eigenfunctions. "It is shown that one single-electron eigenfunction on each of two atoms determines essentially the nature of the electron-pair bond formed between them." He also connects the type of bond to the eigenfunctions involved: "From s and p eigenfunctions, the best bond eigenfunctions which can be made are for equivalent tetrahedral eigenfunctions, giving bonds directed toward the corners of a regular tetrahedron. These account for the chemist's tetrahedral atom and lead directly to free rotation about a single bond but not about a double bond." He also understands that two d eigenfunctions with s and p can give rise to six octahedral eigenfunctions and to the geometry often found in transition metal complexes. In another paper, he focuses on the one-electron and three-electron bonds. He states that "a stable shared-electron bond involving one eigenfunction for each of two atoms can be formed under certain circumstances with either one, two, or three electrons." While any two atoms can form an electron-pair bond, he finds that: "one-electron-bond and three-electron bond, however, can be formed only when a certain criterion involving the nature of the atoms concerned is satisfied." He finds that electron-pair bonds are more stable than the one-electron and three-electron bonds and have roughly twice as large a dissociation energy. He also examines the transition from one extreme bond type to another. In particular, he wonders if there are intermediate bond types between the pure ionic bond and the pure electron-pair covalent bond. He finds that, for instance, alkali halides (CsF, NaCl, LiF, etc.) can be described as essentially ionic. In the case of hydrogen halides (HF, HCl, HBr, HI, etc.), he observes a more complex picture: "The normal HF molecule is represented by neither the formula H^+F^- nor H-F, but by both, with H^+F^- somewhat more important than H-F. The bond is largely ionic." However, for HCl, HBr and HI, "the normal states of these molecules are essentially of the electron-pair bond type, and the formula H-Cl, H-Br and H-I may be used as giving a reasonably accurate picture of the state of the molecules". Finally, he relates the energy of single bonds to the relative electronegativity of the atoms involved. He starts by calculating the energy of covalent bonds between two different atoms, assuming that they are additive. For instance, the energy of a covalent bond between atoms A and B is $E(A-B) = \frac{1}{2}(E(A-A) + E(B-B))$. He then compares this energy to the experimental data and finds that, for ionic bonds, the experimental bond energy far exceeds the energy calculated with the covalent model ($\Delta \gg 0$). On the other hand, he finds a very good agreement between his calculations and the experiments for strongly covalent compound like HI ($\Delta=0$). He relates the difference (Δ) to a new concept named the relative electronegativity of the bonded atoms. "Fluorine and oxygen are by far the most electronegative atoms, with fluorine much more electronegative than oxygen. The series C, N, O, F is almost uniform." He also finds F to be significantly more electronegative than the other halogens and calls fluorine a "super halogen". Moreover, "the amount of ionic character" is related to "the localization of atoms on an electronegativity map". Large electronegativity differences between the two atoms result in strongly ionic bonds and low electronegativity differences lead to strongly covalent bonds. Pauling is awarded the Nobel Prize in chemistry in 1954 "for his research into the nature of the chemical bond and its application to the elucidation of the structure of complex substances". Let us add that he also receives the Nobel Peace Prize in 1962 "for his opposition to weapons of mass destruction".

Semenov (1896–1986) and Hinshelwood (1897–1967) receive the Nobel Prize in 1956 "for their researches into the mechanisms of chemical reactions". Electron transfer research draws more and more interest. Specifically, Taube (1915–2005) (see Figure 4.17) understands that "Simple electron transfer is realized only in systems such as Ne + Ne$^+$. The physics already becomes more complicated when we move to $N_2 + N_2^+$ for example, and with the metal ion complexes which I shall deal with, where a typical reagent is $Ru(NH_3)_6^{2+}$, and where charges trapping by the solvent, as well as within the molecule, must be taken into account, the complexity is much greater." He develops the concept of inner sphere electron transfer for metal complexes in which metal ions are linked to molecules (such as H_2O or NH_3) or ions (such as I^- or Cl^-). He starts by examining the oxidation of a chromium (II) complex, $Cr(H_2O)_6^{2+}$ or $Cr^{2+}(aq)$ into a chromium (III) complex, $Cr(H_2O)_6^{3+}$. To do this, he carries out "a simple test tube experiment", adding solid I_2 to a solution of $Cr^{2+}(aq)$.

FIGURE 4.17   Henry Taube (1915–2005).

He finds that "reaction occurs on mixing, that the product solution is green, and that the green color fades slowly, to produce a color characteristic of $Cr(H_2O)_6^{3+}$. The fading is important because it demonstrates that $(H_2O)_5CrI^{2+}$, which is responsible for the green color, is unstable with respect to $Cr(H_2O)_6^{3+} + I^-$, and thus we could conclude that the Cr(II)-I bond is established before Cr(II) is oxidized." To ascertain this, he carries out another experiment resulting in the formation of the activated complex $[(NH_3)_5Co...Cl...Cr(H_2O)_5]^{4+}$. Inner sphere electron transfer occurs within a complex bridged by an ion Cl^-. Taube demonstrates that "transfer is direct, i.e. Cl^- bridges the two metal centers, and this occurs before Cr^{2+} is oxidized". Taube is awarded the Nobel Prize in Chemistry in 1983 "for his work on the mechanisms of electron transfer reactions, especially metal complexes".

Another class of electron transfer is the outer sphere electron transfer. In 1956, Marcus (1923–) proposes a theory that now bears his name and accounts for electron transfer between molecules in a solution. It is very important since such transfers occur during corrosion and complex chemical reactions, including photosynthesis! Indeed, this theory allows for the determination of the rate of electron transfer reactions or the rate at which an electron jumps from "an electron donor" to "an electron acceptor", such as during the oxidation reaction of Fe^{2+}/Fe^{3+}. It also captures that, in a solution, the electron transfer leads to a rearrangement of charges that results in a polarization of the solvent. Marcus sheds light on the crucial role of the solvent, and, using transition-state theory, determines the activation energy of the reaction and its reaction rate. Marcus also predicts the existence of the "inverted region" in which the electron transfer becomes slower while the reaction is thermodynamically more favorable. This finding is only verified experimentally in 1984 by Miller, Calcaterra and Closs in a study of intramolecular long-distance electron transfer in fluid solutions. Marcus receives the Nobel Prize in chemistry in 1992 "for his contributions to the theory of electron transfer reactions in chemical systems".

In recent years, new discoveries have emerged from photoinduced electron-transfer reactions in heterogenous systems. In particular, the development of artificial photosynthesis (AP) is potentially a huge step forward towards clean energy. The idea is to mimic the natural process of photosynthesis but with a greater efficiency. This may lead to a negative emissions technology in the sense that it would recycle CO_2 to power the planet!

5 Introduction to Chemical Reactivity

Proton Transfer

Als Suchen bedarf das Fragen einer vorgängigen Leitung vom Gesuchten her. Der Sinn von Sein muß uns daher schon in gewisser Weise verfügbar sein. Angedeutet wurde: wir bewegen uns immer schon in einem Seinsverständnis. Aus ihm heraus erwächst die ausdrückliche Frage nach dem Sinn von Sein und die Tendenz zu dessen Begriff. Wir wissen nicht, was »Sein« besagt. Aber schon wenn wir fragen: »was ist ›Sein‹?« halten wir uns in einem Verständnis des »ist«, ohne daß wir begrifflich fixieren könnten, was das »ist« bedeutet. Wir kennen nicht einmal den Horizont, aus dem her wir den Sinn fassen und fixieren sollten. Dieses durchschnittliche und vage Seinsverständnis ist ein Faktum.

Heidegger, "*Sein und Zeit*", 1927

Inquiry, as a kind of seeking, must be guided beforehand by what is sought. So the meaning of Being must already be available to us in some way. As we have intimated, we always conduct our activities in an understanding of Being. Out of this understanding arise both the explicit question of the meaning of Being and the tendency that leads us towards its conception. We do not know what 'Being' means. But even if we ask, 'What is "Being"?', we keep within an understanding of the 'is', though we are unable to fix conceptually what that 'is' signifies. We do not even know the horizon in terms of which that meaning is to be grasped and fixed. But this vague average understanding of Being is still a Fact.

Heidegger, "Being and Time", 1927

EARLY OBSERVATIONS AND CONCEPTS FOR ACIDITY AND BASICITY

Acidity is often associated with food. Indeed, the name acidity comes from the Latin "*acetum*" that, together with wine, gives the famous Latin phrase "*vinum acetum*" which can be translated as "wine turned sour" or vinegar. Let us add that, nowadays, the chemical name for vinegar is acetic acid. As time goes by, cooking evolves from a simple means of preparing food to an art, where ingredients are carefully mixed to obtain new flavors. Just like in chemical reactions! The idea is then to balance our appetite for sugar, fat, acid, salt, etc. For instance, natural sugars in fruits counterbalance their acidity and make them taste better. Acids contained in vinegar, lemon juice, yogurt are often used to tenderize meat and add flavor in a marinade. At the other end of the spectrum, bases or "alkaline" substances (avocados, spinach, broccoli, kale, etc.) are used to neutralize acids, as, for instance, in salads. Most notably, acids and bases are well-known for their properties during baking. Indeed, baking soda (a most common base) reacts with acidic ingredients and causes dough to rise! In a nutshell, when heated, baking soda or sodium bicarbonate ($NaHCO_3$) can react with, for instance, buttermilk to release carbon dioxide gas (CO_2), which gives rise to bubbles in the dough (bread, pizza, cake, etc.). In chemistry, the leavening reaction can be written as $NaHCO_3 + HX \rightarrow NaX + CO_2 + H_2O$, in which HX is an acid and NaX the neutral salt. The same kind of reaction is also used in fire extinguishers that perform very well against different classes of fire. In this case, sulfuric acid (H_2SO_4) is used instead of HX. The chemical reaction is then $2NaHCO_3 + H_2SO_4 \rightarrow Na_2SO_4 + 2CO_2 + 2H_2O$. The idea is to produce a great amount of CO_2 that displaces O_2 and puts out the fire!

An interesting question then arises: how can we distinguish between acids and bases? Boyle finds a first answer when he studies colors. In "Experiments and Considerations Touching Colours" published in 1664, he uses plant decoctions as color indicators. In particular, he observes that blue vegetable juices (berries, cornflowers, etc.) turn red when acids are added, while they become green when bases are added! During the 18th century, Rouelle (1703–1770) uses color indicators such as the *"sirop de violettes"* (violet syrup) to classify salts. Indeed, he notes that an acid combined with another substance yields a neutral salt. He names this mysterious substance a "base". Let us add that in French base means "at the basis of/from". In other words, one needs a "base" and an acid to form a salt. In 1744, in *"Mémoire sur les sels neutres"*, Rouelle proposes a definition for salt: *"Je donne à la famille des sels neutres toute l'extension qu'elle peut avoir: j'appelle sel neutre moyen [...] tout sel formé par l'union de quelqu'acide que ce soit, ou minéral ou végétal, avec un alkali fixe, un alkali volatil, une terre absorbante, une substance métallique, ou une huile."* (I give to this family of neutral salts the broadest form they can have: I call medium neutral salt [...] any salt formed by the union of any kind of acid, either mineral or vegetal, with a solid alkali, gaseous alkali, an absorbent earth, a metallic substance or an oil.) He proposes to rank salts with respect to their acidity: *"J'entends par un sel neutre qui a un excès ou surabondance d'acide, un sel moyen qui outre la juste quantité d'acide qui le met dans l'état neutre parfait, en a encore une nouvelle quantité".* (I mean by a neutral salt, which has an excess or an exceeding amount of acid, a medium salt which, in addition to the exact amount of acid that puts it in the perfect neutral state, contains an extra amount) and with respect to their saturation point. *"J'appelle sel neutre parfait [...] ceux dont le point de saturation est exact [c'est-à-dire] lorsqu'en versant un acide sur un alkali, le movement ou l'effervescence qu'ils ont produit cesse [...] et qui ont une juste quantité d'acide et un degré médiocre de solubilité. [...] ces sels n'altèrent pas le syrop de violettes."* (I name perfect neutral salts those for which the point of saturation is exact [meaning that] when an acid is poured on an alkali, the motion or action they have produced stops [...] and which have the right amount of acid and a mediocre degree of solubility [...] these salts do not discolor the violet syrup.) According to him, there are

FIGURE 5.1   Guillaume-François Rouelle (1703–1770).

many types of salts that can be defined according to their saturation, their acidity and the different crystal forms salts can form. Let us add that Rouelle is a famous apothecary in France (see Figure 5.1). He also organizes scientific demonstrations in his laboratory, drawing large crowds in which one can find Diderot, Parmentier and someone called Lavoisier!

According to Lavoisier, there exists an "acid maker" that he calls oxygen (Greek: ὀξὺς: *oxus* = sour, sharp and γένος: *genos* = birth). This means that, if a substance contains oxygen, it will exhibit acidity! Lavoisier invents a nomenclature that allows the identification of different acids. For example, carbonic acid is composed of oxygen (acid) and carbon (radical or acidifiable base). In other words, carbonic acid is thus the acid associated with carbon. The same is true with phosphoric acid which represents the acid obtained from phosphorus. Moreover, he finds that combustion and acidity are closely related. Indeed, he gathers his findings in "Memoir on the Combustion of Phosphorus Kunckel and the Nature of the Acid that Results from this Combustion" (1777). He states: "Thus, we give the generic name of acids to the products of the combustion or oxygenation of phosphorus, of sulphur, and of charcoal; and these products are respectively named, the phosphoric acid, the sulphuric acid, and the carbonic acid". He also finds that acids "are susceptible of different degrees of saturation with oxygen, and [...] though formed by the union of the same elements, are possessed of different properties, depending upon that difference of proportions [...] When Sulphur is combined with a small proportion of oxygen, it forms, in this first or lower degree of oxygenation, a volatile acid, having a penetrating odour [...] By a larger proportion of oxygen, it is changed into a fixed, heavy acid, without any odour." The first volatile acid is named sulfurous acid while the second heavy acid is called sulfuric acid by Lavoisier. Nevertheless, his nomenclature seems incomplete. For instance, muriatic acid is composed of oxygen and of the muriatic radical. Yet this radical is still unknown!

Everything becomes more complex with the discovery of chlorine by Davy in 1818. Indeed, using electrolysis, he shows that muriatic acid (or hydrochloric acid or spirit of salt) is in fact composed of hydrogen and chlorine (also called oxymuriatic acid)! This means that acids that do not contain oxygen exist! In his experiments, he finds that "One of the singular facts I have observed on this subject, and which I have before referred to, is, that charcoal, even when ignited to whiteness in oxymuriatic or muriatic acid gases, by the Voltaic battery, effects no change in them; if it has been previously freed from hydrogen and moisture by intense ignition in vacuo. This experiment, which I have several times repeated, led me to doubt the existence of oxygen in that substance." It also implies that hydrogen may play a role in acidity, yet, at the time, many hydrogen-containing compounds are known not to be acidic.

Gay-Lussac goes beyond Davy's observations and notes that hydrocyanic acid (HCN) exhibits acidity, yet does not contain any oxygen; the same is true with hydrofluoric acid (HF) and, of course, hydrochloric acid (HCl). He thus creates two categories of acids. The first type consists of oxoacids or acids containing oxygen. The second, called hydracids, are compounds that exhibit acidity, yet do not contain any oxygen. Hydrogen is then deemed central to acidity – what about oxoacids then?

The answer comes in 1838 with Liebig's discovery. Indeed, he shows that, when a metal reacts with an acid, there is production of hydrogen! Here is thus the missing link between hydrogen, acidity and oxoacids! He then proposes a general definition for an acid: "an acid is a compound which contains one or more hydrogens which may be substituted by a metal". In other words, since nothing is created, nothing is lost and everything is transformed, it means that hydrogen is present in the acid when it reacts with a metal since hydrogen is produced at the end of the reaction. The reaction can then be written as: metal + acid (containing hydrogen) → salt + hydrogen (gas). The metal and the non-hydrogen part of the acid combine to form a salt, and the hydrogen part of the acid is retrieved in the hydrogen gas at the end of the reaction. Let us add that Liebig does not propose an explanation for bases. This concept of acidity remains valid until the 19th century and the advent of the first acid–base theory by Arrhenius.

ARRHENIUS AND THE FIRST ACID–BASE THEORY

The first theory that deals with both acids and bases is due to Arrhenius (1859–1927), (see Figure 5.2). He presents most of his ideas in his doctoral dissertation in 1884. His professors are not very impressed by Arrhenius' work and he receives a rather low grade. However, Arrhenius does not stop there and sends his dissertation to other European scientists like Clausius, Ostwald and van't Hoff, who are developing a new field called physical chemistry. Despite being met with opposition at the beginning, his theory becomes extremely important through the efforts of both van't Hoff and Ostwald. At the time, Thomson's ideas on electricity are very influential. Indeed, Arrhenius states: "I was thus led to the assumption that the electrically active

FIGURE 5.2 Svante Arrhenius (1859–1927).

molecules are also chemically active, and that conversely the electrically inactive molecules are also chemically inactive, relatively speaking at least". Arrhenius also makes an important observation: "Concentrated hydrochloric acid, which is free from water, has no effect on oxides or carbonates. It happens that hydrochloric acid in this form is very nearly incapable of conducting electrical current whereas its aqueous solutions have very good conductivity." This means that hydrochloric acid (HCl) is not chemically active unless it is put in solution (mixed with water)! His major idea, developed in his doctoral thesis, is that, in a solution, a compound like HCl dissociates into charged particles or ions. Yet, Faraday's definition of ions only applies to electrolysis, and in Arrhenius' theory, there is no electric current. To better understand his dissociation theory, he proposes that in aqueous solution, hydrochloric acid dissociates into its ions: $HCl\ (aq) \rightarrow H^+ + Cl^-$. Similarly, in water, nitric acid HNO_3 dissociates into H^+ and NO_3^-. Sulfuric acid H_2SO_4 also dissociates into H^+ and HSO_4^-. Perchloric acid $HClO_4$ becomes dissociated in water into H^+ and ClO_4^-. He identifies a common point for all these examples: H^+ is released! He thus gives the following definition for an acid: an acid is a substance that produces an H^+ when put in water (or aqueous solution – Latin: *aqua* = water). Chemically, it can be written as $AH\ (aq) \rightarrow H^+ + A^-$ in which AH is the acid and A^- the corresponding anion after release of H^+. Following the same reasoning for bases, in solution, sodium hydroxide dissociates into two ions according to $NaOH\ (aq) \rightarrow Na^+ + OH^-$. The same is true with potassium hydroxide, KOH (aq) → K⁺ + OH⁻. Similarly, in an aqueous solution, lithium hydroxide dissociates into the cation Li⁺ and the hydroxide anion OH⁻ or LiOH (aq) → Li⁺ + OH⁻. Another example is calcium hydroxide $Ca(OH)_2$ which dissociates into Ca^{2+} and OH⁻ according to the equation: $Ca(OH)_2$ (aq) → Ca^{2+} + 2OH⁻. This leads to the following definition for a base: a base is a substance that produces OH⁻ when put in water (or in aqueous solutions). The corresponding reaction can be written as BOH (aq) → B⁺ + OH⁻ in which BOH is the base and B⁺ the associated cation after release of OH⁻.

Arrhenius then investigates the chemical reaction between an acid and a base. According to him, since both are in solution and produce ions, a chemical reaction takes place between ions! For example, mixing HCl and NaOH in water leads to (H⁺ + Cl⁻) + (Na⁺ + OH⁻) = (Na⁺ + Cl⁻) + HOH. Indeed, in water, HCl dissociates into (H⁺ + Cl⁻) and NaOH into (Na⁺ + OH⁻). He thus finds that this reaction forms a salt (NaCl) dissociated into water as (Na⁺ + Cl⁻) (see a Na⁺ ion in water in Figure 5.3)! In Arrhenius' theory: "salts are in extreme dilution almost completely dissociated into their ions [...] NaCl into Na⁺ and Cl⁻". This reaction is also known as the neutralization reaction of an acid by a base. It can be written as HCl (aq) → H⁺ + Cl⁻ and NaOH (aq) → Na⁺ + OH⁻. Combining both equations, we obtain: HCl (aq) + NaOH (aq) → H⁺ + Cl⁻ + Na⁺ + OH⁻. Then we have: HCl (aq) + NaOH (aq) → NaCl (aq) + H_2O. As we can see, the salt is also dissociated into water as NaCl (aq) → Na⁺ + Cl⁻. Moreover, if we write down the entire equation into its ionic form, we end up with: H⁺ + Cl⁻ + Na⁺ + OH⁻ → Na⁺ + Cl⁻ + H_2O. Let us note that both Na⁺ and Cl⁻ appear on both sides of the reaction, meaning that they act as spectator ions. This leads to a simplified equation of the form:

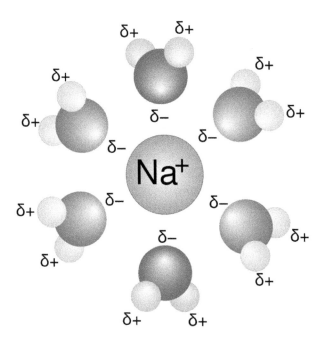

FIGURE 5.3   Sodium ion in an aqueous solution.

$H^+ + OH^- \rightarrow H_2O$. And this is a great discovery! According to Arrhenius: "This equation is equivalent to the formation of water from its two ions, H^+ and OH^- and is evidently independent of the nature of the acid and the base. The development of heat in any reaction of this kind must therefore always be the same for equivalent quantities of any acids and bases". He also finds that this heat of neutralization is always equal to 13,600 calories at 18°C!

Now, let us look at water. Following what precedes, water should dissociate into H^+ and OH^-, or in other words, $H_2O \rightarrow H^+ + OH^-$. This equation is also known as hydrolysis. This means that H_2O is both an acid and a base, since it releases H^+ and OH^-! However, Arrhenius states that "water is hardly dissociated at all" meaning that "water can be regarded as a weak acid or base". He thus introduces the concept of weak acids and weak bases to account for compounds that do not entirely dissociate in water. On the other hand, strong acids or bases are compounds that completely dissociate in water. For example, HCl, HNO_3, H_2SO_4 and $HClO_4$ are strong acids, while $NaOH$, KOH, $LiOH$ and $Ca(OH)_2$ are strong bases. As for weak acids, we have acetic acid H_3COOH which partially dissociates into CH_3COO^- and H^+. The equation is then written as $H_3COOH = CH_3COO^- + H^+$. Here the sign "=" shows the partial dissociation, while the sign "→" is usually used for a complete dissociation. Another example is phosphoric acid H_3PO_4 which gives $H_3PO_4 = H_2PO_4^- + H^+$. Regarding weak bases, we have ammonium hydroxide NH_4OH which dissociates into $NH_4OH = NH_4^+ + OH^-$. Arrhenius receives the 1903 Nobel Prize in Chemistry "in recognition of the extraordinary services he has rendered to the advancement of chemistry by his electrolytic theory of dissociation". During his Nobel lecture, he emphasizes the role of water: "Water, which can be regarded as a weak acid or base, plays an important role. Through its electrolytic dissociation it brings about hydrolysis of the salt of weak acids and bases. It has been possible through observation of the extent of the hydrolysis to calculate the extent of the electrolytic dissociation of water." His findings have important consequences in biology, for instance, for the action of pepsin and trypsin, enzymes that are present in the human digestive tract. Another field of application is geochemistry: "In studying volcanic phenomena, too, the competition between water and silicic acid at different temperatures has proved useful". And finally in the emerging field of catalysis: "the phenomena of catalysis, in which acids or bases play a leading role, have been studied by many investigators, and it has been found that catalytic action depends on the large number of free hydrogen or hydroxyl ions present in solutions".

There are, however, limitations in Arrhenius' theory. Indeed, his theory is based on the presence of water, and it therefore does not account for reactions that take place in other solvents such as benzene (C_6H_6) and chloroform ($CHCl_3$). Also, it does not explain reactions in the gas phase. There is also another drawback to this theory. For instance, ammonia (NH_3) is known to be a base since it discolors color indicators. However, it does not contain OH^- and cannot be classified as an Arrhenius base. Moreover, as we will see, H^+ does not exist on its own in an aqueous solution; it is in fact H_3O^+!

Let us add that Arrhenius participates to the first Solvay Conference in chemistry in 1922 in Brussels, in which "*5 questions d'actualité*" (five timely questions) are discussed: (i) isotopes and radioactivity, (ii) molecular structure and X-rays, (iii) molecular structure and optical activity, (iv) valence and (v) chemical mobility, and someone named T.-M. Lowry also attends the conference.

PROTONATION AND THE BRÖNSTED–LOWRY THEORY

A new acid–base theory emerges in 1923. In fact, two different scientists, Brönsted (1879–1947) in Denmark and Lowry (1874–1936) in England, propose theories that are based on the transfer of protons to explain the acidity and basicity of compounds. On the one hand, Brönsted is interested in Arrhenius' concept of base. Indeed, since ammonia is a base, Arrhenius' theory predicts that it contains OH^- and thus, that its chemical formula should be: $NH_4OH = NH_4^+ + OH^-$. However, the chemical formula for ammonia is known as NH_3. Brönsted thus proposes to write the following equation: $NH_3 + H^+ = NH_4^+$. This is a great development since it means that a base can be defined as a compound able to capture a H^+! There are thus two tremendous advantages to this definition. First, there is no need to specify in which solvent the reaction takes place. Second, an acid can now be defined as the counterpart of a base or, differently put, an acid is a substance that can release a H^+! It opens the door to the possibility of having acid–base couples. In our example, NH_3 is the base and NH_4^+ is its conjugate acid! On the other hand, Lowry is interested in understanding how HCl becomes acidic when mixed with water. In particular, he finds that: "This can be explained by the extreme reluctance of a hydrogen nucleus to lead an isolated existence […] The effect of mixing hydrogen chloride with water is probably to provide an acceptor for the hydrogen nucleus so that the ionization of the acid only involves the transfer of a proton from one octet to another." In other words, this means that $H_2O + HCl = H_3O^+ + Cl^-$. This is an important result since, later on, H^+ is replaced by H_3O^+!

Because of the concomitance of the two proposals and their essentially equivalent conclusions, the overall theory is now called the Brönsted–Lowry theory. "The simplest and most adequate definition of acids and bases is given by the scheme $A = B + H^+$ where A and B represent acid and base respectively. This scheme indicates that an acid is defined as a substance, which is able to split off H^+-ions simultaneously forming a base, and a base as a substance capable of uniting with H^+-ions, thus forming an acid." When the acid–base equilibrium occurs in water or in another solvent, the H^+-ions are solvated. Specifically, in water, $A + H_2O = B + H_3O^+$. This "is a double acid-basic equilibrium, A and H_3O^+ acting as acids, B and H_2O as bases". Another consequence is that acids and bases can either be ions or neutral molecules. For instance, if the acid A is a neutral molecule like formic acid HCOOH, then the base B is an anion called formate anion $HCCO^-$. The chemical equation is then HCOOH $+ H_2O = HCCO^- + H_3O^+$. Indeed, the equation is balanced since there are 4H and 3O on both sides of the equation. The charges are also balanced (the neutrality is ensured on both sides). If the base B is now a neutral molecule such as methylamine CH_3NH_2 then the acid A is, in this case, a cation called methylammonium cation $CH_3NH_3^+$. Again, the chemical equation can be written as $CH_3NH_3^+ + H_2O = CH_3NH_2 + H_3O^+$. As we can see, the numbers of C, N and H are the same on both sides of the equation. Also, there is the same charge (+1) on either side of the equation. This means that, in the Brönsted–Lowry theory, an acid and a base work as a pair or, differently put, as conjugates. Indeed, the conjugate base of an acid is what is left after a proton has been removed from the acid. As an example, the acetate ion CH_3COO^- is the conjugate base of acetic acid CH_3COOH. Similarly, the conjugate acid of a base is what is obtained after adding a proton to the base. The ammonium cation NH_4^+ is thus the conjugate acid of the base ammonia NH_3. There is therefore, according to this theory, an analogy between acid–base pairs, for which one can be obtained from the other by adding or removing a proton, and redox couples, where the oxidizing agent can, for instance, become the reducing agent by accepting electrons. This means that, just like for redox reactions, acid–base reactions can now involve two acid–base conjugates (Figure 5.4). For instance, if the acid A_1 from the A_1/B_1 conjugate pair reacts with the base B_2 from the A_2/B_2 conjugate pair, the overall acid–base reaction is $A_1 + B_2 \rightarrow A_2 + B_1$. In the case of the reaction of acetic acid (CH_3COOH) with ammonia (NH_3), the acid–base pairs involved are CH_3COOH/CH_3COO^- and NH_4^+/NH_3. Such acid–base pairs can be found in tables, like the one introduced by Brönsted in which he also ranks them according to their strengths. In the example above, the acid–base reaction is $CH_3COOH + NH_3 \rightarrow CH_3COO^- + NH_4^+$. In this case, a weak acid reacts with a weak base, which results in a complete ($\rightarrow$) reaction. There can also be incomplete/partial reaction when a weak acid or a weak base reacts with water. For example, $CH_3COOH + H_2O = CH_3COO^- + H_3O^+$. In this case, the two acid–base pairs are CH_3COOH/CH_3COO^- and H_3O^+/H_2O. Water plays here the role of an extremely weak base, leading to an incomplete/partial reaction. Another example is the reaction of a weak base like NH_3 with water, which acts here as a weak acid. The two acid–base pairs are thus NH_4^+/NH_3 and H_2O/OH^-, resulting in the following reaction: $NH_3 + H_2O = NH_4^+ + OH^-$. This is also an incomplete/partial reaction since H_2O is a very weak acid. Finally, the reaction between a strong acid and a weak base will always lead to a complete reaction. For instance, the acid–base pair (H_2SO_4/HSO_4^-) can react with the acid–base pair (CH_3COOH/CH_3COO^-). The acid H_2SO_4 is a strong acid that reacts with the weak base CH_3COO^- to give the complete reaction $H_2SO_4 + CH_3COO^- \rightarrow CH_3COOH + HSO_4^-$.

Acid	Base
HCl	Cl^-
H_2SO_4	HSO_4^-
HNO_3	NO_3^-
H_3O^+	H_2O
HSO_4^-	SO_4^{2-}
H_3PO_4	$H_3PO_4^-$
HF	F^-
CH_3COOH	CH_3COO^-
H_2CO_3	HCO_3^-
H_2S	HS^-
$H_2PO_4^-$	HPO_4^{2-}
NH_4^+	NH_3
HCO_3^-	CO_3^{2-}
HPO_4^{2-}	PO_4^{3-}
H_2O	OH^-
HS^-	S^{2-}
H_2	H^-

FIGURE 5.4   Examples of acid–base conjugates.

A highly significant reaction that relies on the concepts of acids and bases is the saponification reaction (shown in Figure 5.5). It is used, for instance, to manufacture soaps and consists in the reaction of an ester with a strong base like sodium hydroxide (NaOH). An example of an ester is acetyl acetate ($CH_3COOC_2H_5$) which can be formed by reacting a carboxylic acid, here acetic acid (CH_3COOH), with an alcohol which, in this case, is ethanol (C_2H_5OH). The esterification reaction is then written as carboxylic acid + alcohol → ester + water. In our example, $CH_3COOH + C_2H_5OH \rightarrow CH_3COOC_2H_5 + H_2O$. Once the ester is formed, it can be used in a saponification reaction such as $CH_3COOC_2H_5 + NaOH \rightarrow C_2H_5OH + CH_3COONa$. The final product is sodium acetate (CH_3COONa) and is the basis for commercial soap. Alternatively, another example of soap is made from palm oil, $CH_3(CH_2)_{14}COONa$, and is called sodium palmitate. Let us add that the effects of saponification can be seen in oil paintings over time, with soap lumps forming on the surface of the paintings. Indeed, in 2009, the Metropolitan Museum of Art in New York analyzes one of its paintings (Madame X by Sargent), which exhibits an unusual texture and several bumps on its surface. Detailed spectroscopic analyses reveal that the bumps are due to the reaction of a pigment, known as lead white, with the fatty acids from the oil binder. A similar case is observed on a Rembrandt painting ("The Anatomy Lesson of Dr. Nicolaes Tulp") in the Netherlands in 1997. Even if the phenomenon has yet to be fully understood, it has since been found in many paintings created from the 15th century onwards, on canvas, wood or paper. Unfortunately, the alteration due to saponification cannot be reversed, and the only way to restore the painting to its original form is by retouching. The idea is to cover the original paint with varnish and to add new paint, or retouching, over this varnish layer. A second varnish layer is then applied on top of the new paint to stabilize the painting.

A GENERAL APPROACH: LEWIS ACIDS AND BASES

However, Lewis, the inventor of the covalent bond and of the electron dot structure, thinks that acids and bases can be explained through his new formalism. In 1923, he writes in "Valence and the Structure of Atoms and Molecules" that "We are so habituated to the use of water as a solvent, and our data are so largely limited to those obtained in aqueous solutions, that we frequently define an acid or a base as a substance whose aqueous solution gives, respectively, a higher concentration of hydrogen ion or of hydroxide ion than that furnished by pure water. This is a very

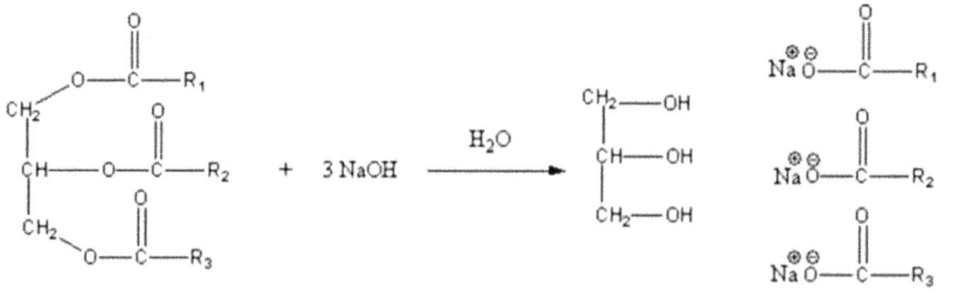

FIGURE 5.5   Saponification reaction.

one-sided definition." He thinks that the definition for an acid or a base should be much more general and be applicable to all solvents. Lewis then gives his own definition of an acid and a base, based on his theory on pairs of electrons: "It seems to me that with complete generality we may say that a basic substance is one which has a lone pair of electrons which may be used to complete the stable group of another atom, and that an acid substance is one which can employ a lone pair from another molecule in completing the stable group of one of its own atoms. In other words, the basic substance furnishes a pair of electrons for a chemical bond, the acid substance accepts such a pair." The power of this new definition is that acids and bases are not defined any longer in terms of hydrogen or hydronium ions, but rather in terms of pairs of electrons and in the possibility for them to create stable groups (duplet and octet rules). This new definition also circumvents the question of the solvent used in the chemical equations, in the sense that, once again, only the pairs of electrons dictate if a substance is acidic or basic. Indeed, compounds with lone pairs of electron substances are basic substances while "On the other hand, substances like hydrogen ion, iodous ion, silicon dioxide, sulfur trioxide and boron trichloride are acid substances since hydrogen ion will accept a pair of electrons to form its stable group of two, and the remaining substances will accept pairs of electrons to complete their stable group of eight." (See Figure 5.6.) For instance, let us look at the reaction between boron trifluoride (BF_3) and ammonia (NH_3). First, start with the Lewis structure of BF_3. At the time, it is well-known (for instance, see Bohr's tables in "Milestones of Modern Chemistry") that B has three valence electrons and F has

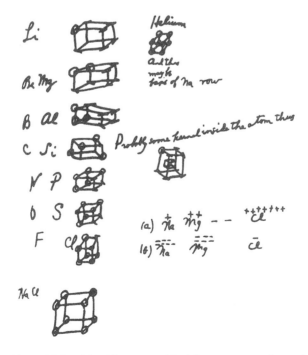

FIGURE 5.6 Gilbert N. Lewis' cubic atoms, with eight corners or valence electrons.

seven valence electrons. This means that, in the BF_3 molecule, there are three simple bonds between B and each of the F, as well as three lone pairs on each F. In other words, boron is surrounded by only six electrons and would need two more to have a complete octet around itself (or a stable group of eight). From Lewis' definition, BF_3 is an acid, since it can accept a pair of electrons (to complete the octet on B). Regarding NH_3, we find from tables from the tables that N has five valence electrons and H has one electron valence. The molecule NH_3 has three single bonds between N and each of the H, with an addition of one lone pair on N. Following Lewis' definition, NH_3 is a base since it can donate an electron pair. We have an acid–base reaction according to Lewis! From an electronic standpoint, the reaction takes place with the lone pair on N moving towards the "empty space" on B (or incomplete valence shell) to form a bond between B and N. This gives the substance ammonia borane (H_3NBF_3). Finally, we write the acid–base reaction as $NH_3 + BF_3 \rightarrow H_3NBF_3$. As we can see from this example, there is no H^+ or OH^- transferred or involved in this reaction. Let us turn to another example: carbon dioxide in water. First, carbon dioxide (CO_2) is composed of one C with four valence electrons and two O with six valence electrons. The molecule CO_2 has thus two double bonds between C and each O as well as two lone pairs on each O. Since CO_2 has four lone pairs, it should be a base. Regarding H_2O, it is composed of two H with one valence electron each as well as one O with six valence electrons. The molecule H_2O has therefore two single bonds and two lone pairs on the oxygen. Since H_2O has two lone pairs, it should also be a base. However, a reaction between two bases does not occur, yet a reaction happens in real life, producing H_2CO_3. Indeed, $CO_2 + H_2O \rightarrow H_2CO_3$. Since the product is H_2CO_3, we can see that a single bond is formed between the C of the CO_2 molecule and the O of the H_2O molecule. This means that H_2O must be the base and CO_2 the acid. So, let us go back to the CO_2 molecule. One way to understand what is happening is to look at the electronegativity of each atom in CO_2. We know that the electronegativity of O is greater than the electronegativity of C. This implies that O draws electrons towards itself, creating polar bonds between C and O. Therefore, C is depleted with electrons, creating "an empty space" that can receive a lone pair from H_2O. This is then the reaction mechanism between the Lewis acid CO_2 and the Lewis base H_2O that forms carbonic acid H_2CO_3. Finally, we can see that electronegativity plays a key role in determining which compound is a Lewis base or a Lewis acid. Most notably, an element that is more electronegative than the other elements can create polarity, or in other words, polar bonds in a molecule. Indeed, electrons belonging to a polar bond will have a tendency to be drawn closer to the most electronegative atom and create a depletion on the least electronegative atom serving in the bond. This least polar side will generally be the locus of an acidic site.

Another application of acid–base reaction is the formation of complex ions. Indeed, the reaction between a transition metal ion and ligands forms what is called a coordination complex (see Figure 5.7). For instance, let us study the formation of a cobalt ammonia complex or hexamminecobalt (III) complex ion: $[Co(NH_3)_6]^{3+}$. This complex is used in X-ray crystallography and NMR to solve the structure of biological molecules including DNA or RNA. Synthesizing this complex implies the use of ammonia (NH_3) and of a cobalt (III) ion (Co^{3+}). In this case, as we have discussed previously, NH_3 is a Lewis base. This means that Co^{3+} is a Lewis acid. The formation of

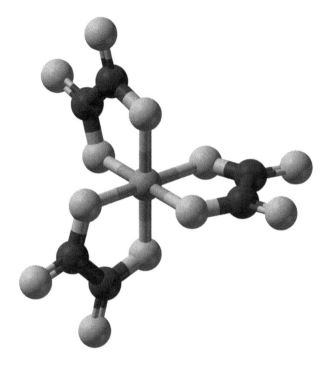

FIGURE 5.7   Iron (III) oxalate ion: here, iron is surrounded by three oxalate groups.

cobalt ammonia complex is then an acid–base reaction that can be written as: Co^{3+} + $6NH_3 \rightarrow [Co(NH_3)_6]^{3+}$. In this reaction, six NH_3 molecules provide their lone pairs to the Co^{3+} ion and form coordinate covalent bonds with the metal. The term coordinate indicates here that the electron pair that forms the covalent bond comes entirely from each of the N atoms of NH_3. This is different from the regular covalent bond seen until now, in which each atom provided one electron to share in the covalent bond. Here the Co^{3+} ion accepts six pairs of electrons or 12 electrons to form the complex. This is due to the fact that, in quantum chemistry, the electronic structure for the Co^{3+} ion is [Ar] $3d^6 \, 4s^0 \, 4p^0$ meaning that there are 12 empty slots (four in 3d, two in 4s and six in 4p) which can accommodate the 12 electrons from the six NH_3. Let us add that the Lewis base NH_3 is called a ligand since it forms a bond with a metal atom.

Metal complexes are found in many biological processes. For instance, the oxygen in the blood is carried by a globular protein called hemoglobin, whose function relies on heme groups or iron (II) complexes. In this case, the ion Fe^{2+} plays the role of a Lewis acid bound to six Lewis bases. These include one nitrogen atom from the globular protein, four nitrogen atoms contained in a heterocycle called porphyrin and one oxygen atom from the O_2 molecule. The iron ion binds reversibly to the oxygen, meaning that it can transport oxygen through the body and release it when necessary. It is the binding between oxygen and iron that gives blood its red color. In 1959, Perutz uses X-ray crystallography to determine the structure of hemoglobin, and receives with Kendrew the 1962 Nobel Prize in chemistry "for their studies on the structures of globular proteins".

THE INTERESTING CASE OF WATER

With the emergence of dissociation theory, Arrhenius and Ostwald want to determine the extent to which a substance is actually dissociated into ions in a solution. To do so, they define the dissociation constant that is equal to the product of the concentrations of the ions divided by the concentration of the undissociated molecule. They note that the value of the dissociation constant can then be used to define the strength of an acid or a base. For instance, in the case of acetic acid, the dissociation equation is given by $CH_3COOH = CH_3COO^- + H^+$. The dissociation constant is also called the acidity constant for an acid. It is calculated as $K_a = [CH_3COO^-][H^+]/[CH_3COOH]$. Ostwald also introduces the dissociation factor α, which corresponds to the fraction of the dissociated acid, and is a number between 0 and 1. For example, $\alpha = 0$ when the acid is not dissociated at all or, differently put, for a dissociation of 0%. For a value of $\alpha = 0.5$, it means that 50% of the compound is dissociated. Finally, when $\alpha = 1$, the substance is completely dissociated (100%). In practice, when a concentration c of CH_3COOH is used at the beginning of the reaction, Ostwald's formalism allows the calculation of the concentration of each substance when equilibrium has been reached. Since a total of αc has been dissociated, it means that we have αc as the concentration for both for CH_3COO^- and H^+ (we had in the beginning a concentration c for CH_3COOH). After reaching equilibrium, $-\alpha c$ has reacted, meaning that, overall, we are left with $c - \alpha c = (1 - \alpha)c$ for CH_3COOH. The dissociation constant K_a is then obtained through $K_a = (\alpha c)(\alpha c)/(1 - \alpha)c$ which can be simplified into $K_a = \alpha^2 c/(1 - \alpha)$. Ostwald notices that "phosphoric, sulphourous, and acetic acids [...] are not ionised beyond about 10 percent under ordinary condition". And he sees that there is "an ionisation of less than 1 percent [for] carbonic acid, sulphuretted, hydrogen, hydrocyanic, silicic, and boracic acids". This means that α is much smaller than 1 and thus $(1 - \alpha)$ is roughly equal to 1. The acidity constant can then be taken to be equal to $K_a = \alpha^2 c$. Therefore, by measuring the concentration of CH_3COO^- at equilibrium, one can obtain the value for αc and thus α. Replacing the values so calculated in the equation $K_a = \alpha^2 c$, we obtain a numerical value for K_a. In this example, $K_a = 1.75 \times 10^{-5}$ for acetic acid.

This framework can also be applied to the dissociation of water. Indeed, Ostwald determines the dissociation constant for the reaction $H_2O = OH^- + H^+$. He calls it K_w and defines it as $K_w = [OH^-][H^+]/[H_2O]$. He finds a value of $K_w = 1 \times 10^{-14}$ in excellent agreement with the value we know today. Ostwald then focuses on the case of bases. He starts on the example of ammonia NH_3. According to him "ammonia dissolves in water to ammonium hydroxide, NH_4OH, which becomes ionised to some extent". This reaction can be written as $NH_3 + H_2O = NH_4^+ + OH^-$. In this case, the dissociation constant is called basicity constant and noted K_b. It is given by $K_b = [NH_4^+][OH^-]/[NH_3][H_2O]$. He finds the value of 2.3×10^{-5} and concludes that "the ionisation amounts to 1.5 percent. Ammonia therefore belongs to the weaker bases". By plugging in the formula for K_w in K_b, we can replace $[OH^-]/[H_2O]$ by the ratio $K_w/[H^+]$ leading to $K_b = [NH_4^+]K_w/[NH_3][H^+]$. Very interestingly, $[NH_4^+]/[NH_3][H^+]$ is the inverse of the acidity constant K_a for the reaction $NH_4^+ = NH_3 + H^+$. This means that K_b is simply $K_b = K_w/K_a$!

Another measure to characterize the acidity or basicity of a substance is introduced by Friedenthal in 1904. The idea is to use different color indicators and match each of them with a specific concentration in H_3O^+ in the solution. He prepares 15 solutions with $[H_3O^+]$ ranging from 1, 1×10^{-1}, 1×10^{-2}, 1×10^{-3}, … to 1×10^{-14} mol/L. Let us add that for 1×10^{-7} he only uses water!

He then works with 14 indicators known as tropaeloine, neutral red, metyl violet, metyl orange, congo red, lackmoid, litmus, galleine, rosolic acid, p-nitrophenol, sodium alizatin, sulfonate, phenolphthalein, naphtholbenzoin and poirrier-blue. After testing the indicators on the $[H_3O^+]$ solutions, he identifies an indicator that changes color for each of the 14 H_3O^+ concentrations. This is the first accurate colorimetric determination of $[H_3O^+]$ in a solution (see Figure 5.8).

Sörensen, a biochemist who works in a brewery in Denmark, comes up with the idea of the "power of H" also known as pH! He writes in 1909: "The value of the hydrogen ion concentration will accordingly be expressed by the hydrogen ion based on the normality factor of the solution used, and this factor will have the form of a negative power of 10". Indeed, he defines the pH as the negative logarithm of $[H^+]$ (in mol/L) or, mathematically, $pH = -\log[H^+]$. $pH = 7$ defines what is called a neutral solution. This is exactly the pH of pure water. When the pH of a solution is less than 7, the solution is acidic, while, when the pH is greater than 7, the solution is basic. Let us add that, in water, the minimum value for a pH is 0 while its maximum value is 14. In other solvents, different from water, it is possible to measure a pH less than 0 for very strong acids and a pH greater than 14 for very strong bases. To illustrate the concept of pH, let us look at the pH for a strong acid such as chlorhydric acid or hydrochloric acid (HCl) with a concentration $[HCl] = 1 \times 10^{-1}$ mol/L. Since HCl is a strong acid, it completely dissociates into H^+ and Cl^- meaning that the concentration in $[H^+]$ is exactly 1×10^{-1} mol/L (as well as $[Cl^-]$). Calculating the pH gives pH $= -\log [H^+] = -\log (1 \times 10^{-1}) = 1$. Similarly, if we have a solution with a concentration

FIGURE 5.8   pH indicator paper.

[HCl] $= 1 \times 10^{-2}$ mol/L, we end up with a concentration in H$^+$ of 1.10^{-2} mol/L. The pH is thus equal to 2. If we look at the example of a weak acid like CH$_3$COOH as discussed previously, we have [H$^+$] $= \alpha c$ with c the initial concentration in CH$_3$COOH. α can be calculated from the acidity constant K$_a$ $= \alpha^2 c$ or equivalently $\alpha = \sqrt{(K_a/c)}$. This means that for c $= 1 \times 10^{-1}$ mol/L, we have $\alpha = 0.013$ and thus [H$^+$] $= 1.3 \times 10^{-3}$ mol/L which, in turn, corresponds to a pH $= 2.9$. In the case where c $= 1 \times 10^{-2}$ mol/L, $\alpha = 0.042$, thus [H$^+$] $= 4.2 \times 10^{-4}$ mol/L corresponding to a pH $= 3.4$. As we can see, for the same concentration, a solution of a weak acid like CH$_3$COOH has a pH greater than a solution of a strong acid like HCl. This is because the dissociation is incomplete in the case of CH$_3$COOH.

Just like the pH is calculated as a negative logarithm, it makes sense to introduce a quantity known as pK$_a$ which represents the "power of acidity" with pK$_a$ $= -$logK$_a$. Going back to the example of CH$_3$COOH/CH$_3$COO$^-$, we have

$$K_a = \left[CH_3COO^- \right]\left[H^+ \right]/\left[CH_3CCOH \right].$$

Since in this case K$_a$ $= 1.75 \times 10^{-5}$, we obtain that the pK$_a$ for the CH$_3$COOH/ CH$_3$COO$^-$ couple is equal to 4.75. If we now take the logarithm of both sides of the equation for K$_a$, we have

$$\log\left(K_a\right) = \log\left(\left[CH_3COO^- \right]/\left[CH_3COOH \right]\right) + \log\left(\left[H^+ \right]\right).$$

Multiplying both sides by -1, we have

$$pK_a = -\log\left(\left[CH_3COO^- \right]/\left[CH_3COOH \right]\right) + pH.$$

This gives the Henderson–Hasselbach equation:

$$pH = pK_a + \log\left(\left[CH_3COO^- \right]/\left[CH_3COOH \right]\right).$$

Similarly, "a power of basicity" called pK$_b$ can be defined as pK$_b$ $= -$logK$_b$. Since K$_b$ $=$ K$_w$/K$_a$, taking the logarithm of both sides gives logK$_b$ $=$ log K$_w$ $-$ log K$_a$. (Recall that $\log(a/b) = \log(a) - \log(b)$). By multiplying both sides by -1, we obtain pK$_b$ $= 14 -$ pK$_a$. Indeed, K$_w$ $= 1 \times 10^{-14}$ and $-\log(1 \times 10^{-14}) = 14$. For instance, for the CH$_3$COOH/ CH$_3$COO$^-$ couple, the pK$_b$ is equal to 9.25.

Finally, a "power of hydroxide" pOH is sometimes used to measure how basic a solution is. It is defined as the negative logarithm of the concentration of hydroxide ions (OH$^-$). The concentration in OH$^-$ can be calculated from the concentration in H$^+$ using the dissociation constant of water K$_w$. Indeed, [OH$^-$] $=$ K$_w$/[H$^+$], since [H$_2$O] is equal by definition to 1, H$_2$O being the solvent, which means that by taking the logarithm of both sides, we have log [OH$^-$] $=$ log (K$_w$) $-$ log ([H$^+$]). Multiplying both sides by -1, we obtain pOH $= 14 -$ pH. This means that, in the case of a solution of a strong acid with a concentration of 1×10^{-1} mol/L, the pH is equal to 1 and the pOH is equal to 13. In the case of pure water, both the pH and the pOH are equal to 7.

TITRATION AND CHEMICAL ANALYSIS

A publication by a French industrial chemist named Descroizilles (1751–1825) marks the start of titration in 1806. In "*Notices sur les alcalis du commerce*" (Notes on Commercial Alkalis), he develops a technique to assess the alkali content in potashes: "Among the American potashes there are three qualities, known as first, second and third grades. The bleaching plants pay for these according to their alkali content". Descroizilles' intuition is that the amount of alkalis in a solution can be determined by adding acids. He then develops an "alkalimeter" to rank these solutions.

Gay-Lussac builds on Descroizilles' work and develops a scientific process to determine accurately the alkali content of a solution. He calls it a titration method (see Figure 5.9). He realizes that the amount of acid poured in Descroizilles' experiments gives the exact amount of base (alkali) contained in the sample! He describes the first titration using a litmus color indicator as "Take the beaker and pour into it the alkaline solution with one pipette [...] pour into it so much litmus solution that should be definitively blue, hold the beaker over a white paper, to perceive better the color change of the litmus. Fill the burette to the mark 0 with the normal acid. Hold the burette in one hand and the beaker in the other, and pour the acid into the alkaline solution, and move the latter constantly in small circles in alternative directions." At this point, he starts examining color changes in the solution: "At the beginning the color does not change but if the potassium carbonate is examined after 11/20 parts of the saturation has taken place then the solution turns to wine red, owing to the liberated carbonic acid". Finally, he concludes "As the saturation point is passed, the wine color of the solution turns to mushroom-red, and the color of the paper turns to red, and remains at this color". In this titration, Gay-Lussac uses as "normal acid" a solution of sulfuric acid (H_2SO_4) which reacts with the "alkaline solution" of potassium carbonate (potash) K_2CO_3. Specifically, H_2SO_4 reacts with the carbonate ions CO_3^{2-} of K_2CO_3 according to the reaction $H_2SO_4 + CO_3^{2-} \rightarrow H_2CO_3 + SO_4^{2-}$. The litmus color indicator (extracted from lichen) is blue when the pH of the solution is above 8.3, and red when the pH gets below 4.5. At the start of titration, the solution of potassium carbonate K_2CO_3 is basic with a pH greater than 8.3 resulting in the blue color for litmus. As sulfuric acid (H_2SO_4) is added to the solution during the titration, the pH of the mixture decreases to finally reach a value below 4.5, giving litmus a "mushroom-red" color.

Another application of acid–base titrations is in the food industry. For instance, titration can be used to test the acidity of orange juice, of wine to determine how it responds to aging or of vinegar as the fermentation process takes place. The taste of vinegar depends directly on the amount of acetic acid in the vinegar. For vinegar to have a good taste, it should contain 4–5% of acetic acid by weight, which corresponds to concentrations between 0.7 and 0.9 mol/L. If there is more acetic acid than 4–5%wt, then the vinegar tastes very sour. Let us examine how the content in acetic acid can be determined by titration of CH_3COOH (see Figure 5.10). First, we take a beaker and put in it 25 mL of vinegar (acetic acid solution) as well as a few drops of the phenolphthalein color indicator. Second, we put 50 mL of a solution of NaOH (with $[NaOH]_t = 0.5$ mol/L) in a burette. The only unknown here is the

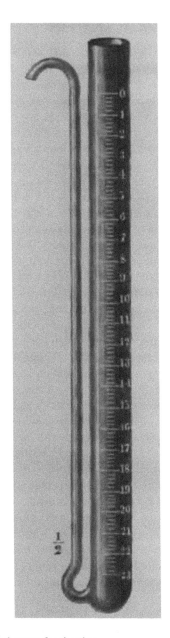

FIGURE 5.9   Gay-Lussac's burette for titration.

actual concentration in CH_3COOH in the vinegar, and the idea is to determine it by monitoring how much of the basic solution of NaOH needs to be added to neutralize CH_3COOH. During titration, the hydroxide ions OH^- react with CH_3COOH and the overall acid–base reaction is $CH_3COOH + OH^- \rightarrow CH_3COO^- + H_2O$. The addition of phenolphthalein allows the visualization of the point at which all of the CH_3COOH has reacted with OH^-. Indeed, this color indicator is colorless if the solution is acidic,

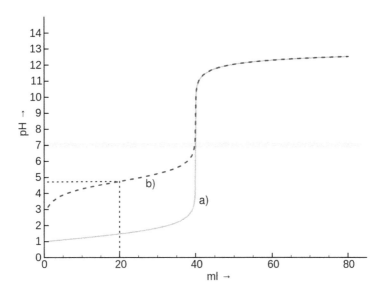

FIGURE 5.10   Titration curve for HCl (a) and for CH_3COOH (b).

neutral or very slightly basic, that is when the pH is less than 8.2. It becomes pink when the pH exceeds 8.2 and the solution becomes basic. Throughout titration, the pH of the solution is measured using a pH meter. Initially, we start from a solution of the acid acetic (25 mL of vinegar) and the pH is equal to 2.4. The phenolphthalein is colorless. As the solution of NaOH is gradually added using the burette, the pH increases very slowly. Its variation can be followed using the Henderson–Hasselbach equation: $pH = pK_a + \log ([CH_3COO^-]/[CH_3COOH])$. Indeed, as NaOH is added, OH^- reacts with CH_3COOH to give CH_3COO^-. This means that the ratio $[CH_3COO^-]/[CH_3COOH]$ increases and, in turn, that the pH increases. After 20 mL of the NaOH solution has been added, the pH is equal to 4.75, which happens to be exactly the value of the pK_a. Taking the Henderson–Hasselbach equation, this means that $\log ([CH_3COO^-]/[CH_3COOH]) = 0$ and thus that $[CH_3COO^-] = [CH_3COOH]$. This means that half of the CH_3COOH initially present has been converted into CH_3COO^- at this point, a stage known as half-equivalent. Then, as more of the NaOH solution is added via the burette, the pH starts to increase more rapidly. When the 40 mL of the NaOH solution has been added, the pH rises rapidly, and the phenolphthalein turns to pink. This means that the solution has become basic, since the entire amount of CH_3COOH present initially has reacted with the added NaOH. At this point, the beaker contains a solution of CH_3COO^-, the conjugate base of acetic acid leading to the basic pH of the solution contained in the beaker. The sharp change in pH and in the color of phenolphthalein gives access to the concentration of CH_3COOH in the vinegar. This stage is known as the equivalence. At the equivalence point, the amount of NaOH added is exactly equal to the amount of CH_3COOH in the vinegar. Mathematically, this is written as $[NaOH]_t \times V_{NaOH} = [CH_3COOH]_{initial} \times V_{CH3COOH}$. Since $[CH_3COOH]_{initial}$ is the unknown, the equation is rearranged as $[CH_3COOH]_{initial} = [NaOH]_t \times V_{NaOH}/V_{CH3COOH}$. Plugging in the data,

$[CH_3COOH]_{initial} = 0.5$ mol/L × 40 mL/25 mL. $[CH_3COOH]_{initial} = 0.8$ mol/L. This means that this vinegar is a good one!

The same method can be applied to determine the concentration of a basic solution. For instance, in the case of a solution of NH_3 with an unknown concentration $[NH_3]_{initial}$, titration involves adding a solution of a strong acid like HCl until the equivalence is reached. In this case, the H^+ added reacts with NH_3 and the titration reaction is written as $NH_3 + H^+ \rightarrow NH_4^+$. At the beginning, we put 40 mL of the NH_3 solution in a beaker and add some methyl red as a color indicator. Methyl red is yellow whenever the pH is above 6.2, and it turns red when the pH gets below 4.4. Initially, the pH of the NH_3 solution in the beaker, as given by a pH-meter, is equal to 11.5 and the methyl red is yellow for such a basic solution. Then, using a burette, we gradually add a solution of HCl with a concentration of $[HCl]_t = 4 \times 10^{-3}$ mol/L. As more and more HCl is added to the beaker, more and more NH_3 gets converted into NH_4^+. As a result, the pH decreases slowly in accord with the Henderson–Hasselbach equation: $pH = pK_a + \log([NH_3]/[NH_4^+])$ since the ratio $[NH_3]/[NH_4^+]$ decreases. After 16 mL has been poured, the pH reaches the value of the pK_a for the NH_3/NH_4^+ couple which is 9.25. At this point, the system has reached half-equivalence, meaning that half of the NH_3 initially present has reacted to give NH_4^+ and that $[NH_3]/[NH_4^+] = 1$. When the added volume of the HCl solution reaches 32 mL, then the pH decreases sharply and the solution turns red. This signals the equivalence. At this point, the NH_3 initially present in the solution has completely reacted with the HCl poured into the solution. This means that the amount of HCl added is exactly equal to the amount of NH_3 initially present and thus that $[HCl]_t \times V_{HCl} = [NH_3]_{initial} \times V_{NH3}$. This leads to the determination of the unknown concentration of NH_3 as 3.2×10^{-3} mol/L.

Maintaining a pH at a given level by adding or removing acids and bases is crucial for living organisms. This is, for instance, the case of the blood pH that, for humans, needs to stay in a very narrow range between 7.35 and 7.45. If the blood pH drops below 7.35, then a condition known as acidosis sets in. If the blood pH becomes greater than 7.45, the corresponding condition is called alkalosis. The key for the organism is to remove the carbon dioxide (or carbonic acid) produced by metabolizing glucose. This is achieved by the lungs which remove CO_2 through breathing, and by the kidneys which remove it through urine. If the blood pH is not in the 7.35–7.45 range, there can be either a respiratory condition or a metabolic issue related to a dysfunction of the kidneys.

RELATIVE STRENGTH AND CHEMICAL HARDNESS

Very early on, Lewis thinks that his theory can be applied to predict how acids and bases react with each other. In 1938, he writes in "Acids and Bases" that "The stronger acids combine with the stronger bases to form stable compounds, while the weaker acids do not usually form compounds with the weaker bases". However, he quickly adds: "Such rough statements regarding the relative strength of acids and bases are useful, but are nevertheless likely to be misleading". As examples for generalized bases, he considers the case of triethylamine $N(C_2H_5)_3$, pyridine C_5H_5N, acetone $(CH_3)_2CO$ and ether $C_2H_5OC_2H_5$, arranged in the order of decreasing basic strength. Similarly, for acids, he studies several examples: sulfur trioxide SO_3, boron trichloride BCl_3, stannic chloride $SnCl_4$, silver perchlorate $AgClO_4$, sulfur dioxide SO_2 and carbon dioxide CO_2, arranged in the order of decreasing acid strength. However, he observes several puzzling facts. For instance, he finds that the silver ion behaves as a weak acid towards the hydroxyde ion. However, adding ammonia to the solution increases the reactivity of the silver ion, which now behaves as a strong acid towards hydroxyde. Furthermore, he finds that weak acids and bases can form stable complexes: "SO_2 is a weak acid and I^- and SCN^- are weak bases but both these anions form fairly stable complexes with SO_2 in aqueous solutions, as can be noticed by the yellow color these complexes impart to the solution". He comes to the overall conclusion that "In a rough way we can understand and sometimes predict this specificity of action between acids and bases", but also that "for any exact formulation far more data are necessary than are now available".

In 1965, Drago and Wayland propose a new approach based on enthalpies of formation of complexes. In "A Double-Scale Equation for Correlating Enthalpies of Lewis Acid-Base Interactions", they introduce the famous equation $-\Delta H = E_A E_B + C_A C_B$ where E_A and C_A are two constants assigned to the acid and E_B and C_B are two constants assigned to the base (see Figure 5.11). They find that "the constants obtained for the acids and bases are interpreted in terms of the electrostatic and covalent nature of the interaction". Indeed, their reasoning is that since the nature of chemical bonding can be understood in terms of ionic and covalent characters, the experimental data for the strength of chemical bonds can be written as a function of ionic and covalent contributions from each of the constituents of the adduct. In other words, when an acid and a base form a complex, the strength of the bonds (of the complex) has two possible origins: E which represents the electrostatic contribution

Base	C_B (kcal./mole)$^{1/2}$	E_B (kcal./mole)$^{1/2}$
C_5H_5N	6.92	0.88
NH_3	3.42	1.34
CH_3NH_2	6.14	1.19
$(C_2H_5)_3N$	11.35	0.65
$HC(O)N(CH_3)_2$	2.73	0.97
$CH_3C(O)OC_2H_5$	2.42	0.639
$CH_3C(O)CH_3$	0.66	0.706
C_6H_6	1.36	0.143

Acid	C_A (kcal./mole)$^{1/2}$	E_A (kcal./mole)$^{1/2}$
I_2	1.000	1.00
C_6H_5OH	0.574	4.70
CH_3OH	0.14	3.41
$(CH_3)_3COH$	0.095	3.77
$HCCl_3$	0.10	5.11
$B(CH_3)_3$	1.76	5.77
SO_2	0.726	1.12
HF	0.0	17.0

FIGURE 5.11 Data for the EC model.

and C the covalent contribution. E_A is then the electrostatic strength brought by the acid, E_B the electrostatic strength from the base, C_A the covalent strength brought by the acid and C_B the covalent strength from the base. For example, for the Lewis acid m-fluorophenol FC_6H_4OH, the parameters are $C_A = 0.506$ and $E_A = 4.42$ while for the base ethylacetate $CH_3COOC_2H_5$, the parameters are $C_B = 1.74$ and $E_B = 0.975$. The two reactants form a hydrogen-bond complex in the solvent carbon tetrachloride (CCl_4). For the complex, the EC model gives $-\Delta H = 5.19$ kcal/mol, in excellent agreement with the experimental calorimetric data of 5.2 kcal/mol. The EC method also allows the comparison of the action of two different bases on the same acid. For instance, if we take the same acid as before, m-fluorophenol FC_6H_4OH, and look at the base pyridine C_6H_5N, $C_B = 6.4$ and $E_B = 1.17$, we have $-\Delta H = 8.41$ kcal/mol. Now, if we consider the base triethylamine $N(C_2H_5)_3$, $C_B = 11.09$ and $E_B = 0.991$, we have $-\Delta H = 9.99$ kcal/mol. Once again, the EC predictions are very close to the experimental data obtained in the solvent cyclohexane (C_6H_{12}). We also see that the $-\Delta H$ for $N(C_2H_5)_3$ takes a greater value than for C_6H_5N suggesting that $N(C_2H_5)_3$ is a stronger base.

Later on, a W term is added to the EC equation leading to the new ECW model for which the equation becomes $-\Delta H = E_A E_B + C_A C_B + W$. This new formulation is important for reactivity and spectroscopy. Indeed, drawing from the ideas of molecular orbital theory, the adduct formation is now interpreted as a charge-controlled reaction, as measured by E, and as a frontier-controlled reaction, as measured by C, while W denotes the intrinsic, constant, contribution from the acid and base. In other words, when the first term ($E_A E_B$) dominates, the reaction is charged-controlled and dominated by electrostatics, while when the second term ($C_A C_B$) dominates, the reaction is frontier-controlled and predominantly covalent. Furthermore, the ECW model can also predict shifts in spectroscopic measurements following complex formation. For instance, during the formation of hydrogen-bond complexes of m-fluorophenol, the infrared stretching frequency of the O–H bond in the phenol shifts. This shift depends on the Lewis base involved in the complex, and its value can be predicted using the E_B and C_B parameters. Let us add that the enthalpies for more than 2,000 complexes have been predicted using the ECW model!

Another theory is introduced in the 1960s by Pearson and is called HSAB for "hard and soft (Lewis) acids and bases". He starts from the reaction between an acid A and a base B written as $A + B \rightarrow AB$. He then states that the equilibrium constant K can be given by a four-parameter equation in the spirit of the EC model: $\log(K) = S_A S_B + \sigma_A \sigma_B$. Here, S_A and S_B represent the strength of the acid and base and σ_A and σ_B the softness of A and B. The value of K is very important for predicting the outcome of the chemical reaction. If $K > 1$, then the formation of the complex AB occurs readily. On the other hand, if $K < 1$, the complex forms in very small amounts. Then, Pearson introduces the HSAB principle in 1965: "Hard acids prefer to associate with hard bases, and soft acids prefer to associate with soft bases". He also defines what hard and soft bases are: Hard bases contain a donor atom that is highly electronegative and has a low polarizability, such as N, O and F. Hard bases are thought of as holding on tightly to their electrons! When the donor atom is P, S, I, Br, Cl or C, Pearson notes that such bases are soft since the donor atoms have a low electronegativity and high polarizability, meaning that soft bases hold on loosely to their electrons! Similarly, a Lewis acid will be soft if it has a large radius and a low positive charge such as Ag^+. On the other hand, a Lewis acid that presents opposite features will be a hard acid such as Mg^{2+}. An application of this principle can be used to determine the sense of a reaction. For instance, let us look at the following reaction: $CH_3Hg^+ + B = CH_3HgB^+$. In this reaction, the methylmercury(I) cation CH_3Hg^+ is a Lewis acid while B is a Lewis base. The complex formation involves the creation of a bond between mercury (Hg) and the base (B). CH_3Hg^+ has a large size, a low positive charge, contains unshared electrons in the p and d orbitals and, as such, is a soft acid. Depending on B, the sense of the reaction will be forward ($\rightarrow$) or $K > 1$ if B is a soft base like I^- or Br^-. On the other hand, the sense of the reaction will be backward ($\leftarrow$) and $K < 1$ if B is a hard base like NH_3, OH^- or F^-. In the latter, very little of the complex will get formed.

In 1983, using computational quantum chemistry, Pearson and Parr introduce the concepts of absolute electronegativity (χ) and absolute hardness (η) (see Figure 5.12). They define $\chi = (I + A)/2$ and $\eta = (I - A)/2$ where I and A are the ionization potential and electron affinity, respectively. A molecule is identified as a Lewis acid or as a Lewis base on the basis of its χ value. Acids are associated with large χ values, while bases have small χ values. A chemical reaction is then interpreted as a transfer of electrons from a molecule of low χ to that of high χ. A high value of η means that an acid or a base is "hard" while a low value of η means high "softness". Pearson and Parr show that when a covalent bond is created between two reactants, the energy lowering ΔE is inversely proportional to the sum of their hardness. Thus, if both the acid and the base are soft, $\eta_A + \eta_B$ is a small number, and ΔE becomes significant and stabilizing: "soft prefers soft". On the other hand, in hard acid–hard base combinations, ionic bonding is favored.

Base	η_B (eV)
F^-	7.0
Cl^-	4.7
H^-	6.8
OH^-	5.6
NO_2^-	4.5
H_2O	7.0
H_2S	5.3
NH_3	6.9

Acid	η_A (eV)
H^+	∞
Na^+	21.1
Ag^+	6.9
Mg^{2+}	32.5
Ca^{2+}	19.7
Cu^{2+}	8.3
Al^{3+}	45.8
CO_2	6.9

FIGURE 5.12   Data for absolute hardness.

CARBANIONS AND CARBON–CARBON BOND FORMATION

During the 19th century, chemists start to experiment by mixing metals with organic compounds. They find that many metals lead to practical difficulties. For instance, alkali metals like sodium (Na) and potassium (K) are very reactive but also very unstable, creating dangerous experimental conditions. Other metals like mercury (Hg) are, on the contrary, too stable and almost do not react with organic compounds. In 1849, Frankland realizes that zinc (Zn) is a good compromise. When mixed with an organic reactant, Zn can be used to form various organic compounds, including hydrocarbons, alcohols and ketones. For example, if we mix zinc (Zn) with methyl iodide (CH_3I), it gives: $2Zn + 2CH_3I \rightarrow Zn(CH_3)_2 + ZnI_2$. Indeed, starting from two molecules of CH_3I, we end up with one molecule (dimethylzinc or DMZ or $Zn(CH_3)_2$) in which Zn is sandwiched between two CH_3 groups. Then, when $Zn(CH_3)_2$ is mixed with water, he finds that "The peculiar behavior of this residue with water, which decomposes it, producing brilliant flame and causing the evolution of pure light carburetted hydrogen (CH_4) induced me to study it more closely". The corresponding chemical equation is: $Zn(CH_3)_2 + 2H_2O \rightarrow 2CH_4 + Zn(OH)_2$. There is production of methane (the shortest and simplest hydrocarbon)! However, the reaction is very slow, taking several weeks to complete, and the yield is very low.

Several scientists follow up on Frankland's work including two well-known Russian chemists Wagner and Saytzeff, as well as a French chemist Barbier. Most notably, Barbier modifies Saytzeff's method by replacing zinc (Zn) with magnesium (Mg). He then assigns to his PhD student Grignard the task of optimizing a chemical reaction (shown in Figure 5.13) or, in other words, increasing the yield (or amount of product obtained). Grignard works for two years on this reaction and completely changes the experimental conditions: "I found that magnesium in the presence of anhydrous ether attacks alkyl halides at ordinary temperature and pressure and that this reaction, which is more or less total, gives a compound which is completely soluble in ether". He realizes that this is a very significant discovery and the birth of a new class of organometallic compounds! Grignard understands that the advantage of magnesium over zinc is that Mg is more electropositive than Zn and has greater reactivity. This means that the reactions previously proposed

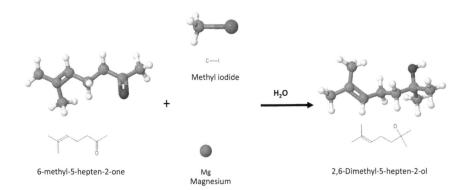

6-methyl-5-hepten-2-one   Mg   2,6-Dimethyl-5-hepten-2-ol
                                Magnesium

FIGURE 5.13   Synthesis of 2,5-dimethyl-5-hepten-2-ol using a Grignard reagent.

with zinc become easier and produce a higher yield when they are carried out with magnesium. Moreover, the synthesis of these organomagnesium compounds is straightforward. The only requirement is that the experiment must be carried out in a dry and inert atmosphere (with no oxygen). Grignard presents his experimental setup: "The apparatus consists of a spherical flask surmounted by a good reflux condenser and a bulb with a tap [...] One gram-atom of magnesium, for instance, is placed in the flask, and one molecule of methyl iodide is dissolved in an equal volume of anhydrous ether. By means of the tap, 25–30 cubic centimeters of this solution are allowed to drop from the bulb onto the magnesium. Almost immediately, there is a very intense reaction, which is moderated by adding, all at once, 200–250 cubic centimeters of anhydrous ether to the flask; it is then maintained by allowing the rest of the reaction mixture to drip on to the magnesium." The reaction is completed very quickly: "All the magnesium disappears, and we finally get a perfectly fluid and colorless liquid". Grignard gives these new compounds the general formula RMgX, in which R is an organic chain (for instance, CH_3 or C_2H_5) and X is a halogen (for instance, Cl or Br) and names them mixed organomagnesium compounds. Nowadays they are known as "Grignard reagents". These reagents are very useful for creating new chemical bonds, and thus new compounds. For instance, RMgX can react with a ketone to form an alcohol. Indeed, if we take RMgX to be CH_3MgI and the ketone to be $(CH_3)_2CO$ (or acetone), we end up with the following reaction $CH_3MgI + (CH_3)_2CO \rightarrow (CH3)_3COMgI$. This product is then put in water (or hydrolyzed) to form t-butanol according to the chemical equation: $(CH_3)_3COMgI + H_3O^+ \rightarrow (CH_3)_3COH + H_2O + Mg^{2+} + I^-$. Such a reaction is at the center of the synthesis of the first local anesthetic known commercially as Stovaine or amylocaine. Indeed, the reaction of ethylmagnesium bromide (C_2H_5MgBr) with monochloracetone ($CH_3(CH_2Cl)CO$) gives an alcohol ($C_2H_5)CH_3C(OH)CH_2Cl$ that is easily transformed into the tertiary amine $(C_2H_5)CH_3C(OH)CH_2N(CH_3)_2$ which is the parent substance of Stovaine. Let us add that this reaction was first discovered by Grignard and Fourneau in 1903. Grignard is awarded the Nobel Prize in chemistry in 1912 "for the discovery of the so-called Grignard reagent, which in recent years has greatly advanced the progress of organic chemistry".

Grignard's discovery draws considerable interest at the beginning of the 20th century. Schlenk is the first to suggest replacing magnesium in the Grignard reagent by lithium, and to synthesize an organolithium compound. To achieve this, he adds lithium metal Li to the organomercury compound diethylmercury $Hg(C_2H_5)_2$ and obtains the organolithium compound LiC_2H_5 known as ethyllithium. This occurs according to the following reaction: $Hg(C_2H_5)_2 + (n+2)Li \rightarrow 2LiC_2H_5 + Li_nHg$. However, these new compounds are highly reactive. As Schlenk writes: "Methyllithium ignites in air and burns with a luminous red flame and a golden-colored shower of sparks". This prompts him to develop new safer apparatus known as "the Schlenk flask", which is a ball-shaped glass container with a nitrogen valve. It is only in 1930 that Ziegler designs a synthesis method that can be scaled up for industrial processes. Organolithium reagents are extremely potent and, as a result, are often used in lieu of Grignard reagents. Their success relies on the same recipe as for organomagnesium compounds, in the sense that the electronegativity of lithium is much lower than for carbon and leads to a highly polarized C–Li chemical bond. This means that, in these organometallic compounds, electrons are drawn towards the carbon. As a result, it creates a quasi-ionic bond between C and Li, that can be written as C^-Li^+. Organolithium is then written as $R_3CLi \rightarrow R_3C^- + Li^+$ with R_3 being three organic chains. The component R_3C^- is then called a carbanion. The presence of a lone pair on C makes carbanions Lewis bases. This means that they can react with Lewis acids according to a specific type of chemical reaction known as nucleophilic addition or substitution. For instance, the attack of the double bond C=O present in ketones or aldehydes can be interpreted as a Lewis acid–base reaction. As an example, methyllithium (CH_3Li) can react with acetone ($(CH_3)_2CO$) to form an alcohol. Here again, CH_3Li can be understood as $CH_3^- + Li^+$ in which CH_3^- is a carbanion, and a Lewis base, that attacks the carbon of the C=O bond in the acetone. The C of the C=O bond reacts as a Lewis acid since it is bonded to a much more electronegative atom (O). This gives the following chemical equation: $CH_3^- + Li^+ + (CH_3)_2CO \rightarrow (CH_3)_3CO^- + Li^+$. The product $(CH_3)_3CO^-$ is then hydrolyzed to give t-butanol $(CH_3)_3COH$. An organolithium compound has also held the record for the strongest existing base. This compound is known as t-butyllithium ($(CH_3)_3CLi$), whose conjugate acid is $(CH_3)_3CH$. Indeed, its pK_a is estimated at 53, which is well beyond what can be measured in water!

Let us add that in 1927 Ziegler discovers that alkalilithiums, for instance RLi compounds, can be added to butadiene ($CH_2=CHCH=CH_2$) or styrene ($C_6H_5CH=CH_2$) to produce polymers known as polybutadiene ($-CH_2-CH=CH-CH_2-)_n$ and polystyrene ($-C(C_6H_5)H-CH_2)_n$. Most notably, polybutadiene is a rubber used to manufacture tires. In 1963, Ziegler, together with Natta, are awarded the Nobel Prize in Chemistry "for their discoveries in the field of the chemistry and technology of high polymers".

CARBOCATIONS: REACTION MECHANISMS AND "MAGIC ACIDS"

In 1901, Norris and Kehrman make a strange discovery. They find that a colorless substance, known as triphenylmethyl alcohol ($(C_6H_5)_3COH$), becomes yellow in solution when mixed with concentrated sulfuric acid (H_2SO_4). This observation puzzles scientists at the time, and Von Baeyer carries out a similar experiment, this time with triphenylmethyl chloride ($(C_6H_5)_3CCl$). He then postulates in 1902 that $(C_6H_5)_3CCl$ is, in fact, an ionic salt-like compound, which dissociates according to the following chemical equation $(C_6H_5)_3CCl = (C_6H_5)_3C^+ + Cl^-$. Von Baeyer also goes on to make the connection between the formation of $(C_6H_5)_3C^+$ and the onset of a color. He calls this process "halochromy" (which changes color with the pH). This phenomenon is often encountered in dyes, which are ubiquitous in nature and give, for instance, their color to red wine, flowers and fruits. Let us add that Von Baeyer receives in 1905 the Nobel Prize in chemistry "in recognition of his services in the advancement of organic chemistry and the chemical industry, through his work on organic dyes and hydroaromatic compounds". However, the existence of this positive ion of carbon compounds, which we now called carbocation, remains an oddity for the next 20 years!

In 1922, Meerwein makes a suggestion that changes the field of organic chemistry. He understands that carbocations can play the role of intermediates during a chemical reaction (see Figure 5.14). Indeed, he finds that the rearrangement of camphene

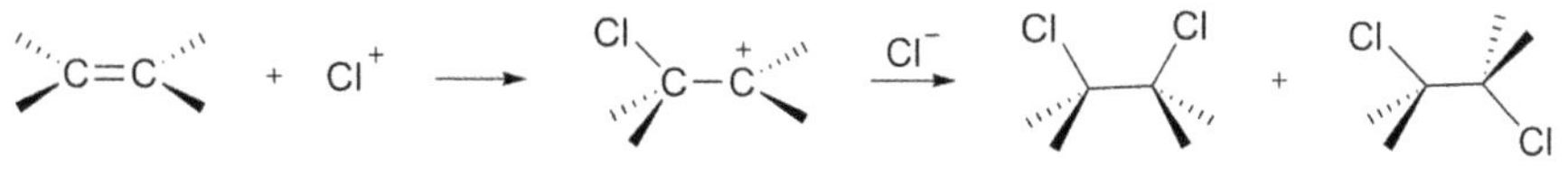

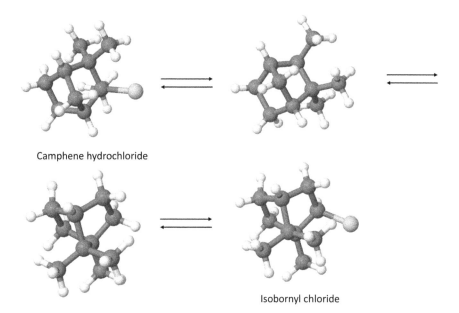

FIGURE 5.14 Example of a carbocation as a reaction intermediate.

hydrochloride to isobornyl chloride is very significantly accelerated by Lewis acids such as antimony pentachloride $SbCl_5$, iron (III) chloride $FeCl_3$ or aluminum chloride $AlCl_3$ (see Figure 5.15). According to him, this means that there must be a cationic intermediate, or carbocation, that forms during the chemical reaction.

A few years later, Ingold and Hughes sense that carbocations may explain a series of chemical reactions, known as nucleophilic substitutions (S_N) and elimination (E) reactions. Nucleophilic substitutions consist of the reaction of an alkyl halide (RX) with a Lewis base (Y) to give a new compound (RY). For instance, the reaction of t-butyl bromide $(CH_3)_3CBr$ with water H_2O gives an alcohol, t-butanol $(CH_3)_3COH$, according to the reaction $(CH_3)_3CBr + H_2O \rightarrow (CH_3)_3COH + HBr$. Here, the Br atom is substituted by the OH group in the product of the reaction. The Lewis base H_2O, through one of the lone pairs on the oxygen, carries out a nucleophilic attack on the carbon **C** in $(CH_3)_3COH$. Ingold and Hughes manage to identify that this type of reaction proceeds according to two successive stages, starting with the dissociation of t-butyl bromide according to the reaction $(CH_3)_3CBr \rightarrow (CH_3)_3C^+ + Br^-$. Then, it is followed by the reaction of the nucleophile H_2O with the carbocation $(CH_3)_3C^+$ to give $(CH_3)_3COH$ according to $(CH_3)_3C^+ + H_2O \rightarrow (CH_3)_3COH + H^+$. They succeed in showing that the reaction rate for this reaction is of order 1 with respect to

Camphene hydrochloride

Isobornyl chloride

FIGURE 5.15 Rearrangement discovered by Hans Meerwein.

$(CH_3)_3CBr$, meaning that it is equal to $v=k$ $[(CH_3)_3CBr]$. In other words, since the reaction rate (v) is proportional to the concentration of $(CH_3)_3CBr$ only, the reaction is said to be unimolecular and is thus called S_N1. In fact, this is always the case! A carbocation will always be an intermediate in an S_N1!

Another example of a reaction with a carbocation as an intermediate is the E1 (unimolecular elimination reaction). For instance, the reaction of t-butyl bromide $(CH_3)_3CBr$ with potassium ethoxide CH_3CH_2OK, which is a strong base, gives 1,1-dimethylethene (an alkene). Indeed, the first stage is again the dissociation of $(CH_3)_3CBr$ to give the carbocation $(CH_3)_3C^+$. The second stage is the reaction of the strong base with the carbocation. The mechanism is the following: the strong base attacks the hydrogen on the carbon next to C^+ (it could be any C from the three CH_3 in $(CH_3)_3C^+$). $CH_3CH_2O^-$ captures this hydrogen and forms an ethanol molecule CH_3CH_2OH, while the carbocation rearranges into the alkene $H_2C=C(CH_3)_2$. The two steps can be written as follows: (5.1) $(CH_3)_3CBr \rightarrow (CH_3)_3C^+ + Br^-$, (5.2) $CH_3CH_2OK + (CH_3)_3C^+ \rightarrow CH_3CH_2OH + H_2C=C(CH_3)_2 + K^+$. In this reaction, there is the formation of a carbon double bond by removing HBr from $(CH_3)_3CBr$. This is the reason why it is called an elimination reaction.

There also exists another reaction family, known as electrophilic substitution, that involves carbocations as intermediates. This is for instance the case of the Friedel and Crafts acylation. An example is the reaction of benzene C_6H_6 with pivaloyl chloride $(CH_3)_3CCOCl$, in the presence of aluminum chloride $AlCl_3$, which gives phenylpivaloyl $C_6H_5COC(CH_3)_3$. In this case, the Lewis acid $AlCl_3$ captures the chlorine atom from $(CH_3)_3CCOCl$ to give the carbocation $[(CH_3)_3CCO]^+$. This intermediate then reacts with the electron-rich benzene molecule to give the product $C_6H_5COC(CH_3)_3$. In summary, the chemical reactions for this electrophilic substitution are: Step (1) $(CH_3)_3CCOCl + AlCl_3 \rightarrow [(CH3)_3CCO]^+ + AlCl_4^-$; Step (2) $C_6H_6 + (CH_3)_3CCO]^+ \rightarrow C_6H_5COC(CH_3)_3 + H^+$.

Let us add that resonance can also play a role in the formation of carbocations. In this case, resonance denotes the possibility of writing different Lewis structures for the same compound by rotating electron pairs. For example, benzene has two resonance (or Kekulé) structures, naphthalene has three resonance structures and so on. Carbocations can also exhibit resonance structures. For instance, triphenylmethyl carbocation has ten resonance structures. This means that this carbocation is particularly stable and thus has a longer lifetime, giving it ample time to react with other reagents involved in a chemical reaction. In such reactions, having a carbocation intermediate makes a lot of sense.

Furthermore, the form of the carbocation itself can play a role during a reaction. Indeed, the inductive effect can be primordial for the realization of S_N1 and E1 reactions. A ternary carbocation R_3C^+ will be much more stable than a secondary carbocation $[R_2CH]^+$, itself more stable than a primary carbocation $[RCH_2]^+$. This means that ternary carbocation will often give rise to S_N1 and E1 reactions, while reactants like RCH_2X will undergo other types of reaction since the carbocation $[RCH_2]^+$ is not stable and thus highly unlikely to form.

The presence of a carbocation as a reaction intermediate has an important consequence for the properties of the products, specifically, for a property known as optical activity or chirality. Indeed, the distribution in space of the substituents R1,

R2, R3 and R4 around a carbon atom in a molecule has a direct effect on its optical activity. In fact, there are two possible distributions (or enantiomers) giving the name of R (Latin: *rectus* = rotation toward the right) or S (Latin: *sinister* = rotation toward the left) to the molecule. When, in a mixture, there are as many R molecules as S molecules, the mixture is called racemic. And this is exactly what happens for the reaction product when a carbocation is a reaction intermediate! In other words, it means that having a carbocation as intermediate will yield a racemic mixture of the products. This is because a carbocation is a planar molecule. Indeed, there are only three bonding pairs around the C^+ (see VSEPR). When a carbocation is a reaction intermediate, there are only two ways for the nucleophilic attack: on the front or on the back of the plane. The distributions of the four substituents will then give rise to either R or S product molecules, but in the end, there will be the same number of R and S molecules! We thus obtain a racemic mixture of product molecules that has no optical activity.

In the 1960s, Olah develops the famous "magic acid" for which he is awarded the Nobel Prize in chemistry in 1994 "for his contribution to carbocation chemistry". "The name magic acid for the $FSO_3H\text{-}SbF_5$ system was given by J. Lukas, a German postdoctoral fellow working with me in Cleveland in the sixties who after a laboratory party put remainders of a Christmas candle into the acid. The candle dissolved and the resulting solution gave an excellent NMR spectrum of the tert-butyl cation. This observation understandably evokes much interest and hence he named the acid 'magic'." Olah shows that it is then possible to create carbocations using superacids! These acids are stronger than mineral acids like highly concentrated sulfuric acid, meaning that the lifespan of a carbocation will be long enough to study it! Indeed, since they are in very acidic systems, they cannot recombine with any bases as these cannot exist! This allows Olah to assemble molecules, disassemble them and even change their structure using carbocations. Indeed, a tremendous advance enabled by superacidic hydrocarbon chemistry is the achievement of efficient low-temperature isomerization reactions of alkanes. This, in turn, opens the door to new methods for methane functionalization, higher hydrocarbons synthesis and environmentally adaptable alkylation reactions. He also adds that "Superacidic systems are not limited to solution chemistry. Solid superacids, possessing both Brönsted and Lewis acid sites, are of increasing significance. They range from supported or intercalated systems, to highly acidic perfluorinated resinsulfonic acids (such as Nafion-H and its analogues), to certain zeolites (such as H-ZSM-5)."

SOLVENT EFFECTS, PROTICITY AND BEYOND

In the second half of the 19th century, chemists start to realize that the rate of chemical reactions depends not only on the reactants, but also the medium, or solvent, in which reactions are carried out. Two French scientists, Berthelot and Péan de Saint Gilles, study the esterification reaction of acetic acid (CH_3COOH) with ethanol (CH_3CH_2OH). Recall that the reaction can be written as $CH_3COOH + CH_3CH_2OH = CH_3COOCH_2CH_3 + H_2O$. In 1862, they publish "*Recherches sur les affinités de formation et de décomposition des éthers*" (Researches on the Affinities of the Formation and Decomposition of Ethers). "When an alcohol is brought into contact with an acid, a combination occurs with varying velocities, depending on the physical conditions of the experiment". They notice that "When the water is removed, the reaction between acid and alcohol can proceed to the end, like the reaction between an acid and a base. In the presence of water, the formation of ether stops at a definite limit. This limit is almost independent of temperature and pressure so long as the system remains liquid." They conclude that the reaction is slowed down in water, meaning that the solvent impacts the way the reaction takes place!

A Russian chemist, Menschutkin, is the first to realize that the solvent is an integral part of the chemical reaction. In his correspondence, he writes about the reaction between trialkyl amines (R_3N) with haloalkanes (RX) in 1890: "*Or l'expérience montre que ces dissolvents exercent sur la vitesse de combinaison une influence considérable. Si nous représentons par 1 la constante de vitesse de la réaction précitée dans l'hexane C_6H_{14}, cette constante pour la même combinaison dans CH_3-CO-C_6H_5 toute chose égale d'ailleurs sera 847,7. La différence est énorme, mais, dans ce cas, elle n'atteint pas encore le maximum.*" (Yet, the experiment shows that these solvents have a considerable influence on the combination rate. If we give a value of 1 to the rate constant of the reaction mentioned above in hexane C_6H_{14}, this constant will be 847.7 for the same combination in CH_3-CO-C_6H_5, all other things being equal. The difference is huge, but, in this case, it hasn't reached yet its maximum.) He then concludes "*Vous voyez que les dissolvents, soit disant indifférent ne sont pas inertes, ils modifient profondément l'acte de la combinaison chimique. Cet énoncé est riche en conséquence pour la théorie chimique des dissolutions*". (You can see that the solvents supposedly indifferent are not inert, they deeply change the act of chemical combination. This statement is full of consequences for the chemical theory of dissolution.) In other words, solvents participate in chemical reactions and their physical and chemical properties have a direct impact on the rate of chemical reactions!

In 1887, Wislicenus discovers that "Ethyl, formyl, phenyl acetate can exist under two forms. These are two isomers of the same molecule, or in other words different molecular structures for the same molecule, which can easily rearrange each into the other" (see Figure 5.16). The form on the right is called enol while the form on the left is called keto, and the chemical equilibrium between the two is known as keto–enol equilibrium or tautomerism. Wislicenus observes that the keto-form is more soluble in sodium carbonate (Na_2CO_3) solutions than the enol-form. This prompts him to study the chemical equilibrium between the two forms in a series of solvents. He goes on to show that the ratio between the two forms depends on the solvent, with the keto-form being predominant when alcohols (ROH) are used as solvent, while the

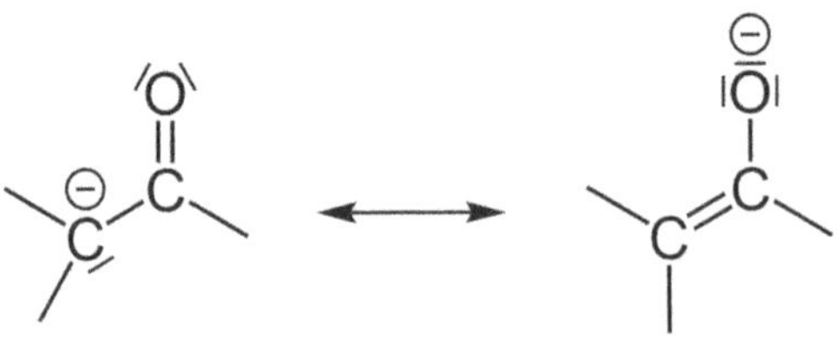

FIGURE 5.16 Principle of the keto–enol isomerism or tautomerism.

enol-form is predominant in benzene (C_6H_6) or chloroform ($CHCl_3$). This discovery shows that, in some cases, solvents can change the value of the equilibrium constant or, differently put, the ratio according to which reactants and products are obtained at the end of the chemical reaction.

This leads to a classification of solvents according to their properties. Two main features are important: polarity and proticity. Polarity depends on the polarization of the chemical bonds within the molecule. This comes from the difference in electronegativity between atoms in the molecule. For instance, water, acetone and chloroform are all polar solvents. On the other hand, benzene, cyclohexane and hexane are non-polar solvents. Proticity denotes the ability of a solvent to form hydrogen bonds with reactants. For this purpose, a protic solvent must contain a hydrogen bonded to a highly electronegative atom like O or N. Examples include water, ammonia and ethanol. On the other hand, an aprotic solvent cannot form such hydrogen bonds. This is, for instance, the case for acetone, benzene and cyclohexane. Protic solvents can also give off H^+ in an acid–base reaction. A monoprotic acid can release $1H^+$ as in acetic acid, a diprotic acid can release $2H^+$ as in H_2SO_4 and a triprotic acid can release $3H^+$ as in H_3PO_4. Polar protic solvents are often used to dissolve salts and play a key role in S_N1 reactions. This is because they create conditions that favor the dissociation of the alkali halide (RX) reactant. Indeed, they stabilize the carbocation intermediate and thus increase the rate of unimolecular nucleophilic substitution. On the other hand, polar aprotic solvents do not stabilize the carbocation intermediate and do not allow for S_N1 reactions to take place. Instead, another type of reaction (or mechanism) known as bimolecular nucleophilic substitution (S_N2) occurs. This type of reaction bypasses the formation of a carbocation intermediate and achieves the substitution in a single step.

In recent years, the emphasis on developing a sustainable chemistry has led to the development of a new class of solvents, called neoteric solvents, which are environmentally benign. Such solvents include ionic liquids, supercritical fluids and perfluorohydrocarbons. Furthermore, the use of water in organic synthesis and the development of solvent-free reactions on solid mineral supports like clays have drawn considerable interest in recent years.

The concept of proticity is also used in biology. Indeed, Mitchell (1920–1992) introduces proticity as: "I use the word proticity for the force and flow of the protons current by analogy with the word electricity, which describes the force and flow of an electron current". He receives the Nobel Prize in chemistry in 1978 "for his contribution to the understanding of biological energy transfer through the formulation of the chemiosmotic theory". He presents his theory in 1961 in "Coupling of Phosphorylation to Electron and Hydrogen Transfer by a Chemiosmotic Type of Mechanism". Indeed, he finds that "the processes that we call metabolism and transport represent events in a sequence, not only can metabolism be the cause of transport, but also transport can be the cause of metabolism." According to him, there are two fundamental principles that rule the bioenergetically efficient mechanisms. First, the lipid membrane and plug-through complexes play the role of an osmotic barrier and separate the proton conductors. Second, plug-through complexes catalyze conduction of electrons, hydrogen atoms and ions, as well as oxide anions. For example, specific ligand binding include the electron-accepting action of cytochromes or ironsulfur centres, the hydrogen-accepting action of flavoproteins or Q-proteins, and the O^{2-} accepting action of the ATP/(ADP$^-$ + P$^-$) couple. Finally, he adds that "I think it is fair to say that the protonmotive property of the mitochondrial cytochrome system and the photosystems of chloroplasts can probably be correctly explained, in general principle, by the direct ligand-conduction type of chemiosmotic mechanism".

6 Introduction to Inorganic Chemistry
Role of Symmetry

Tous les événements sont enchaînés dans le meilleur des mondes possibles: car enfin si vous n'aviez pas été chassé d'un beau château à grands coups de pied dans le derrière pour l'amour de mademoiselle Cunégonde, si vous n'aviez pas été mis à l'Inquisition, si vous n'aviez pas couru l'Amérique à pied, si vous n'aviez pas donné un bon coup d'épée au baron, si vous n'aviez par perdu tous vos moutons du bon pays d'Eldorado, vous ne mangeriez pas ici des cédrats confits et des pistaches. – Cela est bien dit, répondit Candide, mais il faut cultiver notre jardin.

Voltaire, "Candide ou l'Optimisme", 1759

"There is a concatenation of all events in the best of possible worlds; for, in short, had you not been kicked out of a fine castle for the love of Miss Cunegund; had you not been put into the Inquisition; had you not traveled over America on foot; had you not run the Baron through the body; and had you not lost all your sheep, which you brought from the good country of El Dorado, you would not have been here to eat preserved citrons and pistachio nuts." "Excellently observed," answered Candide; "but let us cultivate our garden."

Voltaire, "Candide or Optimism", 1759

SYMMETRY, AESTHETICS AND GOLDEN RATIO

Since the antiquity and the ancient Greeks, the question of aesthetics (Greek αίσθητικόσ: *aisthetikos* = of/for perception by the senses) or, in other words, the definition and recognition of beauty is essential as it pertains to one's system of values. Indeed, according to Plato in *"Philebus"* (4th century BC), there is a link between measure, symmetry and beauty: "And now the power of the good has retired into the region of the beautiful; for measure and symmetry are beauty and virtue all the world over". Specifically, he states how to find the good: "Then, if we are not able to hunt the good with one idea only, with three we may catch our prey; Beauty, Symmetry, Truth are the three, and these taken together we may regard as the single cause of the mixture, and the mixture has being good by reason of the infusion of them". According to Aristotle, beauty is also related to structure. In *"Poetics"*, he writes "To be beautiful, a living creature, and every whole made up of parts must […]

present a certain order in its arrangement of parts". He goes further in metaphysics and explains the role played by mathematics: "The chief forms of beauty are order and symmetry and definiteness, which the mathematical sciences demonstrate in a special degree".

It seems that numerous architectural monuments are designed with symmetry in mind, so as to give them a beautiful aspect. For instance, the Great Pyramid of Giza (around 2560 BC) shows a ratio of the base to the height of 1.5717 and a ratio of the circumference to the height of 6.28 (very close to 2π). The Parthenon in Athens (5th century BC) is another example, with a ratio of width to height of 2.25 and a ratio of length to width of 2.25 as well. In the following centuries, many artists use rules of proportions. Among then, we find Da Vinci and his famous "Vitruvian man" describing the perfect proportions for the human body (see Figure 6.1). The

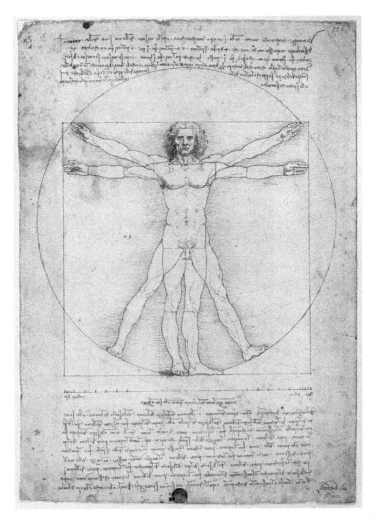

FIGURE 6.1 Da Vinci's Vitruvian man (16th century).

mathematical formulation of beauty is first stated in 1509 in *"Divina proportione"* by Pacioli. Let us add that this book is illustrated by Da Vinci! Especially, Pacioli describes the divine proportion, that we can nowadays write according to the following equality between ratios, i.e. $a/b = (a+b)/a$. For instance, if $a = 10$ and $b = 8$, $a/b = 10/8 = 1.25$ and $(a+b)/a = 18/10 = 1.8$. Since 1.25 is not equal to 1.8, the rectangle defined with sides a and b is not in "divine proportion". But if $a = 5$, $b = 3$, $a/b = 5/3 = 1.67$ and $(a+b)/a = 8/5 = 1.6$, it means that this rectangle is in "divine proportion". Such calculations are used in numerous masterpieces in art, such as in the Mona Lisa by Da Vinci, and in paintings by Raphael, Michelangelo, Rembrandt and even Seurat, Mondrian and Dali during the 20th century! But, let us add that 3 and 5 are also the first two successive Fibonacci numbers!

In 1202, Fibonacci introduces his famous "Fibonacci sequence" in *"Liber abbaci"*. He begins with the following problem "How many pairs of rabbits can be bred from one pair in one year? A man has one pair of rabbits at a certain place entirely surrounded by a wall. We wish to know how many pairs can be bred from it in one year, if the nature of these rabbits is such that they breed every month one other pair, and begin to breed in the second month after their birth." The results are as follows: "Let the first pair breed a pair in the first month, then duplicate it and there will be two pairs in a month. From these pairs, one, namely the first breeds a pair on the second month, and thus there are three pairs in the second month. From these, in one month two will become pregnant, so that in the third month two pairs of rabbits will be born. Thus there are five pairs in this month." After six months, there are 21 pairs, and after one year, 377. "And this number of pairs has been born from the first-mentioned pair at the given place in one year." Fibonacci then starts to understand how the sequence of numbers is built: "You can see in the margin how we have done this, namely by combining the first number with the second, hence 1 and 2, and the second with the third, and the third with the forth. At last, we combine the 10th with the 11th hence 144 with 233 and we have the sum of the above-mentioned rabbits, namely 377, and in this way you can do it for the case of infinite numbers of months." Nowadays, this sequence of numbers is represented as $F_0 = 0$, $F_1 = 1$ and $F_n = F_{n-2} + F_{n-1}$ for $n > 1$ (with F for Fibonacci). This means that we have the sequence F_0, F_1, F_2, F_3, F_4 ... or, equivalently, 0, 1, 1, 2, 3, 5, 8, 13, 21, 34, 55, 89, 144, 233, 377 ... This sequence can also be calculated geometrically as the sum of squares as in Figure 6.2.

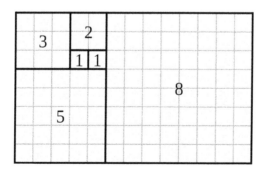

FIGURE 6.2 Fibonacci's blocks.

Let us add that Fibonacci is, in fact, Leonardo da Pisa, another important Leonardo in Italy! Now, if we look at the ratio of two successive Fibonacci numbers, we find something very surprising. For instance, $F_2/F_1 = 1$, $F_5/F_4 = 1.67$, $F_{11}/F_{10} = 1.6181818$, $F_{14}/F_{13} = 1.618026$, $F_{21}/F_{20} = 1.6180339$. After reaching $n = 27$ or F_{27}/F_{26}, this ratio equals 1.618033989 which is equal, up to its nine decimal places, to the golden ratio φ. Let us add that φ is also the solution of the quadratic equation $x^2 - x - 1 = 0$ with a value of $\varphi = (1 + \sqrt{5})/2 = 1.6180339887$. In 1812, Binet introduces a formula, known as Binet's formula, that describes Fibonacci's numbers in closed form. Indeed, this formula allows the direct calculation of the nth Fibonacci number using a recurrence relation. It is written as $u_n = u_{n-1} + u_{n-2}$ for $n > 1$, with $u_0 = 0$, $u_1 = 1$ and $u_n = [(1 + \sqrt{5})^n - (1 - \sqrt{5})^n]/[(2^n \sqrt{5}]$. Finally, let us add that the term "φ" (phi) is coined by Barr at the beginning of the 20th century, supposedly in honor of Phidias, the Greek sculptor who worked on the Parthenon in Athens. φ also appears in Penrose's work on tiles in the 1970s. His study elucidates how a plane can be covered entirely with two types of tiles, that never leave a gap in between them and never overlap with each other. In particular, he identifies tiles with a five-fold symmetry that look like either "darts" or "kites", whose sides must have a ratio of φ for the plane to be completely covered! Penrose is also a famous mathematical physicist, who works with Hawking on the theory of general relativity and elucidates the physics of black holes leading to the well-known Penrose–Hawking singularity theorems. In the 1980s, φ appears in a new form of matter known as quasicrystal. In 1982, Mackay obtains a tenfold diffraction patterns by putting circles (representing atoms) at the intersections of Penrose's mosaic. At the same time, Shechtman uses electron microscopy and finds that there actually exists a tenfold symmetry in real life. Indeed he observes something very strange! Atoms in a crystal are generally packed in patterns that are repeated over and over in the three directions of space. However, Shechtman's image does not correspond to what is expected: it shows regular patterns that are aperiodic. These patterns follow mathematical rules but cannot be repeated over and over again. Nowadays, scientists describe quasicrystals using the golden ratio. Indeed, "the ratio of various distances between atoms is related to the golden mean". In 1984, Steinhardt and Levine make the connection between the theoretical model of Mackay and the experiment of Shechtman, meaning that Penrose tiling can explain Shechtman peculiar crystals! In 2011, Shechtman is awarded the Nobel Prize in chemistry "for the discovery of quasicrystals".

Finally, let us add that symmetry is everywhere, in human activities as well as in nature. Indeed, the number of petals in flowers follows the Fibonacci sequence. For instance, lilies have three petals, buttercups have five, cosmos have eight, asters have 13, while the spoon mum species of chrysanthemum has 21! DNA is sometimes seen as another example, as each turn of the double helix is 21 angstroms wide and 34 angstroms long, giving a ratio of 1.619 (again in the Fibonacci sequence, 34 and 21 are successive numbers). We also find the use of the golden ratio in architecture. Specifically, Le Corbusier thinks that order, the golden ratio and the Fibonacci sequence are related: "Geometry is the language of man [...] he has discovered rhythms, rhythms apparent to the eye and clear in their relations with one another. And these rhythms are at the very root of human activities. They resound in man by an organic inevitability, the same fine inevitability which causes the tracing out of the Golden Section." Symmetry also appears in music in which the golden ratio is used in Bartok, Bach, Satie and Debussy's work. In chemistry, symmetry is an important concept in quantum mechanics, spectroscopy and crystallography, which are based on the mathematical principles found in group theory.

NUMBER THEORY, GROUP THEORY AND GROUP REPRESENTATIONS

The development of group theory during the 20th century changes how we envision symmetry and its implications in various disciplines such as chemistry. This new mathematical formalism relies on number theory, algebraic equations and geometry.

Number theory is a field that primarily studies integers. As early as the 6th century BC, Pythagoras investigates integer triples, such as (a,b,c) that gives $a^2 + b^2 = c^2$. These are known as Pythagorean triples and form the basis for the famous Pythagorean theorem, which states that, in a right triangle, the square of the hypotenuse is equal to the sum of the squares of the other two sides. For instance, the triple (3,4,5) is the most well-known example as $3^2 + 4^2 = 9 + 16 = 25 = 5^2$. Other examples include (8,15,17) and (7,24,25). Later, Euclid starts to study prime numbers in his book "Elements" and states the following definitions on even and odd numbers: "6. An even number is that which is divisible into two equal parts. 7. An odd number is that which is not divisible into two equal parts, or that which differs by a unit from an even number. 8. An even-times-even number is that which is measured by an even number according to an even number. 9. An even-times-odd number is that which is measured by an even number according to an odd number. 10. An odd-times-odd number is that which is measured by an odd number according to an odd number.". He also defines prime numbers such as: "11. A prime number is that which is measured by a unit alone. 13. A composite number is that which is measured by some number" as well as their properties in Book 9: "Proposition 20: Prime numbers are more than any assigned multitude of prime numbers" or, differently put, there is an infinity of prime numbers! For instance, 3 is a prime number, since it cannot be divided by any number other than 1. The same is true for 5, 7, 11 ... These are the first principles for number theory. During the 17th century, Fermat (1607–1665) works on the so-called "perfect" numbers and "amicable" numbers. Indeed, a perfect number is equal to the sum of its aliquot divisors, or divisors that are less than the number itself. For instance, 6 has aliquot divisors 1, 2 and 3 and since we have $1 + 2 + 3 = 6$, it means that 6 is a perfect number! Also, two numbers are said to be amicable if the first equals the sum of the aliquot divisors of the other, and vice versa. For instance, 220 and 284 are amicable since 220 has the following aliquot divisors: 1, 2, 4, 5, 10, 11, 20, 22, 44, 55, 110. And we have $1 + 2 + 4 + 5 + 10 + 11 + 20 + 22 + 44 + 55 + 110 = 284$. Similarly, 284 has aliquot divisors 1, 2, 4, 71, 142 and we have $1 + 2 + 4 + 71 + 142 = 220$. Thus 220 and 284 are amicable numbers! Moreover, Fermat is extremely interested in prime numbers. In 1640, he proposes the famous "Fermat's little theorem": "*Tout nombre premier mesure infailliblement une des puissances −1 de quelque progression que ce soit, et l'exposant de la dite puissance est sous-multiple du nombre premier donné -1; et, après qu'on a trouvé la première puissance qui satisfait à la question, toutes celles dont les exposants*

sont multiples de l'exposant de la première satisfont tout de même à la question [...]
Et cette proposition est généralement vraie en toutes progressions et en tous nom-
bres premiers; de quoi je vous envoierois la démonstration, si je n'appréhendois
d'être trop long." A more modern statement is that if p is a prime number, then for
any integer a, the number $a^p - a$ is an integer multiple of p. For example, if $a = 4$ and
$p = 5$, then $4^5 = 1024$ and $1024 - 4 = 1020 = 5 \times 204$ is an integer multiple of 5. He
also challenges the entire mathematical community with his famous "Fermat's last
theorem" or "Fermat's conjecture" (1637). This conjecture remains one of the most
difficult mathematical problems until the proof is found more than three centuries
later in 1995! The statement in Latin is the following: "Cubum autem in duos cubos,
aut quadratoquadratum in duos quadratoquadratos, et generaliter nullam in infini-
tum ultra quadratum potestatum in duos ejusdem nominis fas est dividere: cujes rei
demonstrationem mirabilem sane detexi. Hanc marginis exiquitas non caperet". In
other words, there are no nonzero integers a, b, c, n with $n > 2$ such that $a^n + b^n = c^n$.
For centuries, finding the proof for this theorem continues to be impossible until
Wiles' proof in 1995. Indeed, Wiles receives the Abel Prize in 2016 "for his stunning
proof of Fermat's Last Theorem by way of the modularity conjecture for semistable
elliptic curves, opening a new era in number theory" and the 2017 Copley Medal by
the Royal Society "for his beautiful and unexpected proof of Fermat's last theorem
which is one of the most important mathematical achievements of the 20th century".

Another mathematician, Euler (1707–1783) is interested in Fermat's work.
One of his most impressive achievements is the proof that the sum of the recipro-
cals of the primes diverges. $\sum_{p \text{ prime}} 1/p = 1/2 + 1/3 + 1/5 + 1/7 + 1/11 + \ldots = \infty$. Let
us add that Euler is also known for Euler's number "e" which is approximately
equal to $e = 2.71828$ (the famous relation ln e = 1). Later, Gauss builds on Euler's
work and develops modular arithmetic in *"Disquisitiones Arithmeticae"* in 1801.
To better understand this new concept, let us focus on an example. If we think
of what time it is during the day, we already have a first instance of modular
arithmetic. Indeed, looking at a 12-hour clock, we can see that it is 4 o'clock
twice during the day (4 a.m. and 4 p.m.). This is because, on such a clock, the
day is basically divided into two periods of 12 hours each. In this case, we say
that the time of the day is defined by an integer "modulo" 12. More generally, in
modular arithmetic, there is a congruence relation between the integers defined
as follows: for a positive integer n, two numbers a and b are said to be congruent
modulo n, if their difference $a - b$ is an integer multiple of n, also denoted by $a \equiv b$
(mod n). For example, $43 \equiv 3$ (mod 8) since $43 - 3 = 40 = 5 \times 8$. In addition, $-19 \equiv 2$
(mod 7) because $-19 - 2 = -21 = 3 \times 7$. A contemporary application is found in
identification numbers, such as the ISBN number for a book, the VIN (Vehicle
Identification Number) for a car or bar codes. For example, take the ISBN-13
978-0-367-20828-8. The sum of products is $9 \times 1 + 7 \times 3 + 8 \times 1 + 0 \times 3 + 3 \times$
$1 + 6 \times 3 + 7 \times 1 + 2 \times 3 + 0 \times 1 + 8 \times 3 + 2 \times 1 + 8 \times 3 = 122$ We find that 122
(mod 10) = 120 remainder 2. 2 is not equal to 0 but $10 - 2 = 8$, which is exactly
the value of the "check digit", the 13th digit of the ISBN-13 number! This means
that the ISBN is valid (our book!).

Yet the power of group theory is far more impressive! Indeed, the question of
solving quintic equations (polynomial equations of degree 5) still remains a very

challenging problem at the end of the 18th century. Lagrange in *"Sur la résolution algébrique des équations"* (1771) understands that new methods need to be developed to solve these equations. Furthermore, Abel (1802–1829) confirms: *"Au lieu de demander une relation dont on ne sait pas si elle existe ou non, il faut demander si une telle relation est en effet possible [...] de telle ou telle manière"*. Lagrange understands that symmetry and permutation play a major role in finding the solutions of polynomial equations. A young French mathematician Galois (1811–1832) (see Figure 6.3) finally finds the solution in 1831 but dies one year later during a duel. His idea is to define a new method based on "groups": *"On peut se donner arbitrairement une première permutation, pourvu que les autres permutations s'en déduisent toujours par les mêmes substitutions de lettres. Le nouveau groupe ainsi forme jouira évidement des mêmes propriétés que le premier."* (One may take an arbitrary first permutation, if the others are deduced from the same substitutions of letters. The newly obtained group will have the same properties as the first one.) For instance, if we take the sequence "A B C D" then the permutations give "B A D C", "C D A B" and "D C B A" giving us a group of four elements. Regarding the quintic problem, Galois theorizes that the group keeping all the qth roots $p^{1/q}$, in which q is the prime of any polynomial p, invariant has at most $1/(q(q-1))$ times as many elements as the group leaving p invariant. Other mathematicians develop group theory, including Cauchy (1789–1857), Cayley (1821–1895) and Lie (1842–1899). In particular, Cauchy introduces the theory of permutation groups, as well as the theory of matrices. Klein (1849–1925) in his 1872 Erlangen program classifies geometries by their symmetry groups, linking geometry and group theory. But the most impressive work comes from Frobenius (1849–1917) and his idea of representation theory. Thanks to this new formalism, it is possible to simplify complex problems in terms of group representations and their associated characters. More specifically, the elements of a group can be described as matrices and the characters as the traces of matrices. During the 20th century, Weyl (1885–1955) and Wigner (1902–1995) make crucial contributions by linking group theory and quantum mechanics! According to Wigner, symmetry, and especially group theory, is the key: "I have come to agree [...] that the recognition that almost all rules of spectroscopy follow from the symmetry of the problem is the most remarkable result". In 1963, Wigner receives the Nobel Prize in physics "for his contributions to the theory of the atomic nucleus and the elementary particles, particularly through the discovery and application of fundamental symmetry principles". Bethe (1906–2005),

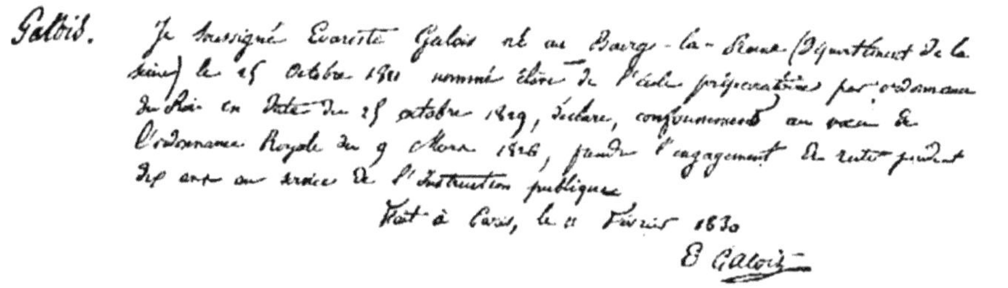

FIGURE 6.3 Evariste Galois' ten-year commitment to public education.

another Nobel Prize awardee, shows that group theory can be applied to problems concerning the nature of crystals.

Let us add that one of the most famous applications of group theory can be seen in the Rubik's Cube. Rubik (1944–), a Hungarian architect, invents this cube to help students better understand structures in 3D. His idea is that the person "playing with this cube" can "experience something of the pure logic of the universe, of its boundless essence, of its perpetual motion in space and time [...] to get to know the inexhaustible richness of symmetries in space." In the 1980s, mathematicians are interested in using group theory to solve the cube. However, no study has shown yet how to determine with certainty the minimum number of moves needed to solve the cube. To this day, one only needs 26 quarter-turn moves!

X-RAYS AND THE ADVENT OF CRYSTALLOGRAPHY

The study of stones has always been of interest to scientists. They start with the external appearance and classify stones according to their shapes, weights, etc. However, a discovery by Abbé Haüy (1743–1822) changes everything. Indeed, in 1784, he writes in *"Essai d'une théorie sur la structure des cristaux"*: *"Une obfervation que je fis sur le fpath calcaire en prifm a fix pans, terminé par deux faces exagones, me fuggera l'idée fondamentale de toute la théorie dont il s'agit".* (An observation I made on a prismatic calcite crystal with six sides, and terminated by two hexagonal sides, suggested to me the fundamental idea for the theory I devised.) Indeed, he drops this crystal and finds that it becomes cleaved along one of the edges of the base." *J'avois remarqué qu'un criftal de cette variété, qui s'étoit détaché par hazard d'un groupe, feu trouvoit caffé obliquement, de manière que la fracture préfentoit une coupe nette, & qui avoit ce brillant auquel on reconnoit le poli de la Nature."* (I had remarked that a crystal of this type, that got detached by chance from a group, was broken obliquely, so that the fracture showed a neat cut, and had a very shiny aspect in which one can recognize the polish from Nature.) Pursuing his investigations, he manages to divide this crystal into smaller pieces until he detaches a rhomboid nucleus. He then takes other stones and repeats his experiment. Every single time he concludes: *"J'y retrouvais le même noyau rhomboidal que m'avoit offert le prifm dont j'ai parlé plus haut".* (I recovered the same rhomboid nucleus that I had obtained from the prism I discussed earlier.) Haüy understands that crystals are periodic structures and are all composed of small polyhedral, like this rhomboid nucleus, that he calls the *"molécule intégrante".*

Frankenheim and Hessel see the importance of symmetry in crystals. Indeed, considering symmetry operations such as rotations, translations, etc., they find that there should exist 32 possible crystal classes. However, Bravais (1811–1863), (see Figure 6.4), a French physicist, develops a new theory on the *"structure réticulaire* of crystals" in 1849. According to him, it is possible to link the internal symmetry of a crystal with its external shape! He concludes that one only needs to define 14 space lattices, now known as the Bravais lattices, to characterize all crystals present in nature. In particular, he examines: *"les divers genres de symétrie que peut offrir une molécule crystalline, considerée comme un système d'atomes, et representée par un*

FIGURE 6.4   Auguste Bravais (1811–1863).

polyhèdre dont ses atomes occupent les sommets". (The various types of symmetry that can be found in a crystalline molecule, considered as a system of atoms, and represented by a polyhedron in which the atoms are located on the corners.) Then, he identifies: *"les lois suivant lesquelles la symétrie de la molécule se transmet en partie au système réticulaire, formé par les centres de gravité des diverses molécules dont un crystal se compose"*. (The laws according to which the symmetry of the molecule is propagated in part to the reticular system, formed by the centers of mass of the various molecules that compose the crystal.)

A new scientific field appears with the advent of X-rays at the beginning of the 20th century. Its name is crystallography and its goal is to study crystal structures using X-rays. In 1912, von Laue (1879–1960) discovers that, when X-rays are projected on a crystal of copper sulfate, there is a diffraction pattern that appears on a photographic plate! In 1914, he receives the Nobel Prize in physics "for his discovery of the diffraction of X-rays by crystals". Von Laue works in Munich (Germany) where Röntgen (1901 Nobel Prize in physics) develops X-rays and where another well-known physicist, Sommerfeld (who holds the record for most Nobel nominations with the number of 84!), is also heavily involved in X-ray research. Von Laue states in his Nobel Lecture that "Since the times of Haüy and Bravais the basic crystallographic law of rational indices had been explained simply and visually by the mineralogists through space-lattice arrangement of the atoms. Sohncke, Federow and Schonflies had brought the mathematical theory of possible space-lattices to the greatest possible degree of perfection. But no more far-reaching physical conclusion had evolved from this line of thought." During a conversation with Ewald (1888–1985), von Laue understands that X-rays can prove the existence of crystal lattices. Indeed, "during the conversation I was suddenly struck by the obvious question of the behavior of waves which are short by comparison with the lattice-constants of the space-lattice. And it was at that point that my intuition for optics suddenly gave me the answer: lattice spectra would have to ensue." Indeed, the lattice constant in a crystal is of the order 10^{-8} cm while the wavelength of X-rays is around 10^{-9} cm! "Thus, the ratio of wavelengths and lattice-constants was extremely favorable if X-rays were to be transmitted through a crystal. I immediately told Ewald that I anticipated the

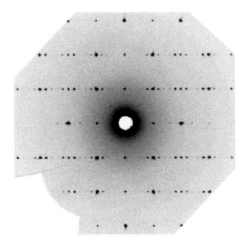

FIGURE 6.5   Example of an X-ray diffraction pattern.

occurrence of interference phenomena with X-rays." He carries out the following experiment: "Copper sulfate serves as the crystal [...]. The irradiation direction was left to chance. Immediately from the outset the photographic plate located behind the crystal betrayed the presence of a considerable number of deflected rays [...]. These were the lattice spectra that had been anticipated." X-ray crystallography was born! (See Figure 6.5.)

After discussing von Laue's work with his father (William Henry, 1862–1942), William Lawrence Bragg (1890–1971) derives a formula, known today as Bragg's law, that deciphers X-ray diffraction patterns. He starts from von Laue's experiments on a crystal of zinc sulfide and tries to understand why, "whereas there were a large number of directions in which one would expect to find a diffracted beam, only a certain number of these appeared on the photographic plate used to record the effect". Bragg posits that "perhaps we ought to look for the origin of this selection of certain directions of diffraction in the peculiarities of the crystal structure". He therefore interprets the diffraction patterns as the reflection of the X-rays by the planes that define the crystal structures. Indeed, according to him, "the points of a space lattice may be arranged in series of planes, parallel and equidistant from each other". This leads him to propose the famous Bragg's law as $n\lambda = 2d\sin\theta$, in which n is an integer, λ denotes the wavelength of the X-rays, d represents the spacing of the planes in the crystal and θ the angle according to which the X-rays are reflected. This leads Bragg to identify correctly the structures of a wide range of crystals. For instance, in the case of the zinc blende structure, it is assumed that von Laue interprets the diffraction pattern as being produced a simple cubic lattice, while Bragg's analysis shows that it is in fact a face-centered cubic lattice as now known. Similarly, he studies the X-rays photographs of sodium chloride and potassium chloride. While these two compounds are very similar and are expected to have the same crystal structure, von Laue's photographs indicate that sodium chloride is a face-centered lattice, while potassium chloride looks like "one would expect from an arrangement of points at the corners of cubes". Bragg has the explanation: atoms are arranged in a way such that "every corner of the cube is occupied by an atom, whereas the atoms of one kind considered alone are arranged on the face centered lattice". The structure for the two compounds is exactly the same, and the difference between the two diffraction patterns can be explained as follows: "in potassium chloride the atoms are so nearly equal in their weights that they act as equivalent diffracting centers and the structure may be regarded as a simple cubic one". For their achievements, father and son receive the Nobel Prize in Physics in 1915 "for their services in the analysis of crystal structure by means of X-rays". Let us add that in 1915, Lawrence is 25 years old, making him the youngest ever Nobel laureate in Physics!

However, the story of X-ray crystallography does not end there. In fact, numerous advances are made using this technique in biology. Indeed, Dorothy Crowfoot Hodgkin (1910–1994) receives the Nobel Prize in Chemistry in 1964 "for her determinations by X-rays technique of the structures of important biochemical substances", including the structure of penicillin in 1946 and the structure of vitamin B_{12} in 1956. Let us add that vitamin B_{12} is known to have the most complex structure of all vitamins! Another important achievement due to the use of X-ray crystallography is the discovery of the DNA diffraction picture known as "Photo 51", showing the helix structure of DNA, by Rosalind Franklin (1920–1958). In 1962, Watson (1928–), Crick (1916–2004) and Wilkins (1916–2004) receive the Nobel Prize in Physiology or Medicine "for their discoveries concerning the molecular structure of nucleic acids and its significance for information transfer in living materials", four years after Franklin's death.

MOLECULAR SYMMETRY, SPECTROSCOPY AND CRYSTAL STRUCTURE

After several decades of intense research, group theory finally identifies that there are 230 space groups. To better explain the different concepts of point groups, type of lattices, crystallographic point groups and space groups, let us consider two examples. Let us start with the water molecule and identify the molecular point group of H_2O. First, the H_2O molecule is nonlinear. Second, the rotation of the molecule by $180°$ (or $360°/2$) leaves the molecule unchanged (this is called a symmetry operation C_2). There are also two more symmetry operations (mirrors planes – see Figure 6.6) that leave the molecule unchanged: σ_{xz} (plane that contains the molecule) and σ_{yz} (plane perpendicular to the molecule). H_2O is thus associated with the point group C_{2v}. Now that we have this information, let us look at the character table for the point group C_{2v} to determine some of the properties of H_2O (see Figure 6.7). First, we can see that the operations of symmetry discussed above are summarized in the first line: E (identity), C_2, σ_{xz}, σ_{yz}. Then, the first column includes the labels A_1, A_2, B_1 and B_2 which are called the symmetry representations. The labels A or B are attributed on the basis of the effect of the rotation about the principal axis. If the representation is symmetric (unchanged), then it is called A. Otherwise, if it is antisymmetric (reversed), it is called B. Let us add that Mulliken proposes a standardization of notations for polyatomic molecules. To do so, he sends his system of notation to about 150 molecular spectroscopists and scientists in related fields for their comments and suggestions, and refines the notations based on their responses. In the case of the C_{2v}

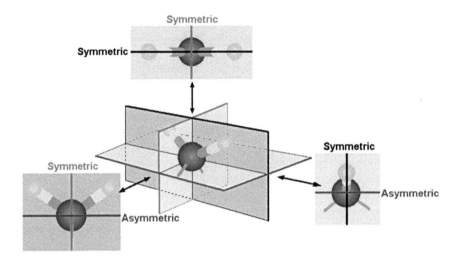

FIGURE 6.6 Mirror planes for the water molecule.

C_{2v}	E	C_2	$\sigma_v(xz)$	$\sigma_v'(yz)$		
A_1	1	1	1	1	z	x^2, y^2, z^2
A_2	1	1	-1	-1	R_z	xy
B_1	1	-1	1	-1	x, R_y	xz
B_2	1	-1	-1	1	y, R_x	yz

FIGURE 6.7 Character table for H_2O.

group, he identifies that there is some confusion between the representations B_1 and B_2 which are "distinguished by their behavior for reflection in the 2 planes σ_{xz} and σ_{yz}". He notes that "the z-axis is the axis of symmetry, but the choice of the x- and y-axes, hence the identification of the 2 planes is arbitrary, with 2 choices possible". To clarify this, Mulliken recommends that "for planar C_{2v} molecules, the x-axis always be chosen perpendicular to the plane of the molecule". The +1 and −1 in the table correspond to the symmetric or antisymmetric character, respectively. On the right-hand side, the last two columns include information about various functions. Depending on the character in the table (+1 or −1), one can deduce what the change operated by the operations of symmetry on these functions will be. Now if we look at functions centered on the oxygen atom, the behavior of these functions is given directly by the table (in terms of characters), since the C_2 axis goes through the oxygen atom and the planes contain the O atom of H_2O. For instance, if we take a function depending on z, such as, for example, the p_z orbital on the oxygen atom, we can see that all the symmetry operations E, C_2, σ_{xz}, σ_{yz} have a character of +1 in the table, meaning that these symmetry operations keep the orbital unchanged or, in other words, that the orbital is symmetric. The representation A_1 also applies for a function depending on x^2, y^2 or z^2, similar to a sphere that stands for the s orbitals of the O atom. For a function depending on x only, such as for example the p_x orbital of O, we can see from the third row of the table that both E and σ_{xz} leave the orbital unchanged (+1), while the operations C_2 and σ_{yz} lead to the opposite behavior (or antisymmetric) for the orbital.

Indeed, the positive and negative lobes are swapped after the application of C_2 and σ_{yz}. The character table also provides info on the R_x, R_y or R_z rotations. Indeed, if the rotation along the z-axis (R_z) is subjected to the symmetry operations E and C_2, then the resulting rotation still occurs in the counterclockwise sense, meaning that it remains the same rotation (symmetric). On the other hand, when R_z is subjected to the operations σ_{xz} and σ_{yz}, then the resulting rotation takes place in the opposite sense, the clockwise sense (antisymmetric). The same is true with R_x and R_y. Other functions can also be studied using the character table. These are for instance functions dependent on z^2 as in the d_{z2} orbital of the O atom, or dependent on the product xy as in the d_{xy} orbital. In summary, the orbitals of the O atom can be A_1, B_1, A_2, B_2. Now let us turn to the orbitals on the H atoms of H_2O. When we apply C_2 to the two s orbitals of hydrogen, we see that the s orbital of the first H (on the left, or H1) is transformed into the s orbital on the second H (on the right, or H2). It is neither symmetric nor antisymmetric. The idea is therefore to propose a linear combination of both s orbitals. For instance, a first orbital can be noted as s(H1) + s(H2) while the second orbital can be noted s(H1) − s(H2). When C_2 is applied to both orbitals, we see that the first one stays unchanged meaning that the combination s(H1) + s(H2) is symmetric, while the second combination s(H1) − s(H2) is antisymmetric. Applying the same reasoning to σ_{xz} and σ_{yz}, we see that the first s(H1) + s(H2) has an A_1 representation (sequence: 1, 1, 1, 1), while the second one s(H1) − s(H2) has a B_2 representation (sequence: 1, −1, −1, 1). In summary, the atomic orbitals on the oxygen atom (only accounting for the s and p orbitals) are A_1, B_1, B_2 and the atomic orbitals on the hydrogen atoms (only accounting for the s orbitals) are A_1 or B_2.

Now if we are interested in the bond symmetry in H_2O, one needs to look at the reducible representation Γ_σ that characterizes the σ bond in H_2O. Let us add that the same procedure can be applied to other types of bonds such as the π bonds, represented by Γ_π. Indeed, the use of the character table shows how bonds are affected by symmetry operations. So, if we look at Γ_σ for water, there are two σ bonds. These bonds can be regarded as vectors. The E operation leaves everything unchanged, so the character is 2 (1 + 1) since there are two bonds, with each of them contributing +1, since the operation leaves both bonds unchanged. The C_2 operation moves the two bonds giving a character 0 (0 + 0) for this symmetry operation. The σ_{xz} operation moves the two bonds giving a character equal to 0 (0 + 0), while σ_{yz} leaves the bonds unchanged giving a character equal to 2 (1 + 1). We thus have the following sequence for the characters associated with E, C_2, σ_{xz}, σ_{yz}: 2, 0, 0, 2. To get the irreducible Γ_σ, we need to find a linear combination of the symmetry representations A_1, B_1, A_2 and B_2 to be equal to the sequence 2, 0, 0, 2. And there is only a single possibility: $\Gamma_\sigma = A_1 + B_2$. Indeed, when we add the characters for A_1 and B_2, we have the following sequence (1 + 1), (1 − 1), (1 − 1), (1 + 1) or 2, 0, 0, 2!

This means that in order to create a σ bond, A_1 and B_2 must be combined. It involves mixing orbitals on both the O and H atoms. In other words, a σ bond can be created by mixing an atomic orbital of O of the A_1 representation with the combination of H orbitals that has an A_1 representation (they are symmetrically compatible). The same is true with the B_2 representations for both the atomic orbital on O and the combination of H orbitals that is B_2. More specifically, obtaining the A_1 representation, or the first σ bond, relies on mixing of the 2s orbital on O with the s(H1) + s(H2) combination for the H atoms. It is also possible, from a symmetry standpoint, that a $2p_z$ orbital on O mixes with the s(H1) + s(H2) combination on the H atoms. Another possibility is the mixing of a combination of the 2s orbital and $2p_z$ orbital on O with the s(H1) + s(H2) combination on the two Hs. For the B_2 representation, the $2p_y$ orbital on O is mixed with the s(H1) – s(H2) combination on the H atoms.

When looking at the results from quantum mechanics (QM) calculations using computers, we indeed find that the 2s orbital on the O atom mixes with a combination of the two s orbitals on the H atoms, leading to a bonding molecular orbital $2a_1$. As a convention, minuscule letters are used for molecular orbitals. This constitutes the first σ bond. Then, the $2p_y$ orbital on the O atom mixes with the s(H1) – s(H2) combination on the H atoms to give a bonding molecular orbital $1b_2$. This constitutes the second σ bond. The next molecular orbital is formed by mixing a combination of 2s and $2p_z$ on O with the s(H1) + s(H2) combination on the Hs. This gives the molecular orbital $3a_1$ that is essentially nonbonding. This shows that group theory is a very powerful tool, since it gives a pretty good approximation of molecular bonding within the H_2O molecule by simply considering the symmetry of the molecule and by just using a pen!

It is also possible just by using symmetry considerations to deduce the type of vibrations and if they are observable by infrared spectroscopy, as well as by Raman spectroscopy. Indeed, in the case of water, we find that $\Gamma_{vib} = 2A_1 + B_2$ meaning that water has three distinct vibrations! More symmetry-based calculations show that the A_1 and B_2 vibrations are observable by both infrared spectroscopy and by Raman spectroscopy.

Finally, a last application of group theory is the study of crystal structures. For instance, let us have a look at the crystal of water, known as ice. The most common form of ice is known as hexagonal ice (Ih). The crystal structure of Ih is defined by the space group $P6_3/mmc$ (number 194) and its crystallographic point group is D_{6h}. Ice Ih is the form of ice found in nature, for instance in snow. Let us add that are many other crystal structures for ice that can be obtained at high pressure or very low temperature, leading to the 18 different ice crystals known today!

CRYSTAL FIELD THEORY, MAGNETISM AND COLORS

One of the great mysteries that remains unexplained at the start of the 1950s is the behavior of complexes of transition metals, especially their magnetic properties and the many colors they can take. The beginning of an answer comes from the work of two physicists Bethe and van Vleck. Both receive a Nobel Prize in physics, Bethe in 1967 "for his contributions to the theory of nuclear reactions, especially his discoveries concerning the energy production in stars", and van Vleck in 1977 "for [his] fundamental theoretical investigations of the electronic structure of magnetic and disordered systems". van Vleck recalls that: "In 1930 I held a Guggenheim fellowship for study and travel in Europe. I spent most of the time in Germany, but by far the most rewarding part of the trip scientifically was a walk which I took with Kramers along one of the canals near Utrecht." Among other topics, they discuss Kramers' work as well as "Bethe's long paper concerned with the application of group theory to the determination of the quantum mechanical energy levels of atoms or ions exposed to a crystalline electric field". After returning to the US, he works with his research group on the development of crystal field theory. He then applies it to understand the magnetic properties of salts of rare earth metals and of metals from the iron group. As explained by van Vleck, "the basic idea of the crystalline field potential is an extremely simple one". Indeed, he believes that the metal at the center of the complex can be considered as an ion with a positive charge surrounded by ligands with a negative charge. These ligands thus create a static electric field "which is regarded as an approximate representation of the forces exerted upon [the metal ion] by other atoms in the crystal". He also recognizes the significance of symmetry: the form of the crystalline potential depends on the type of crystalline symmetry!

During the 1950s, Orgel understands how the crystal field or electrostatic theory can be applied to understand the electronic structure and colors of metal transition compounds. Indeed, he takes the example of the $[Cr(NH_3)_6]^{3+}$ complex ion, which can be treated as a chromium (III) ion Cr^{3+} in a crystal field of six ammonia molecules (NH_3). Here, NH_3 is a neutral molecule that serves as a ligand. However, it bears an unshared electron pair on the N atom and in the complex ion, the negative end of the lone-pair dipole always points towards the metal. This means that the ligands give rise to a field equivalent to that created by a set of negative charges located around the metal ion. Specifically, for transition metal complexes, the crystal field or electrostatic field created by the ligands has a direct impact on the energies of the d orbitals of the metal ion. This means that the electrostatic field splits the d orbitals and that instead of having five d orbitals with the same energy, there can be for instance two groups of orbitals with, say, three and two orbitals respectively that have different energies. Let us start with the example of a metal ion complex with six ligands as in $[Cr(NH_3)_6]^{3+}$. From VSEPR, we know that the geometry of this complex is octahedral (AB_6). To understand the effect of the crystal field, we need to look at two things: first, the geometry of each of the d orbitals and second, the location of the ligands in space. Starting with the d_{z^2} orbital in an octahedral complex, the lobes of the orbitals along the z-axis point towards the two NH_3 on top and at the bottom of the complex, meaning that there is a strong repulsion between the electrons of the d_{z^2} orbital and the lone pairs of the NH_3 ligand. The same is true with $d_{x^2-y^2}$ which points towards the four NH_3 ligands in the central plane. This means that, overall, these two orbitals have a higher energy as a result of the strong electronic repulsion. On the other hand, the three other d orbitals d_{xy}, d_{yz} and d_{zx} have their lobes pointing in between the ligands and thus experience little electronic repulsion from the ligands. As a result, the five d orbitals from the isolated metal ion become split into two groups of orbitals, a first group composed of two orbitals (d_{z^2} and $d_{x^2-y^2}$) of higher energy and a second group with three orbitals (d_{xy}, d_{yz} and d_{zx}) of lower energy. According to group theory, our complex belongs to the O_h point group. The two orbitals (d_{z^2} and $d_{x^2-y^2}$) fall into the symmetry representation E_g (g: German: *gerade* = even or symmetric with respect of the inversion operation), while the three orbitals (d_{xy}, d_{yz} and d_{zx}) correspond to the T_{2g} symmetry operation. In terms of electron configuration, Cr^{3+} has three valence electrons in its d orbitals. When Cr^{3+} is surrounding by six NH_3 ligands, the electrons occupy the lowest energy orbitals and thus the electron configuration can be written as $(t_{2g})^3 (e_g)^0$.

The effect of the crystal field on the d orbitals strongly depends on the geometry of the complex. For instance, if we consider a tetrahedral complex, then the splitting occurs in a completely different way. In this case, none of the d orbitals point exactly towards the four ligands. However, the orbitals that point most closely to the ligand electrons are the three orbitals (d_{xy}, d_{yz} and d_{zx}) while the other two (d_{z^2} and $d_{x^2-y^2}$) point further away from the electrons. The electronic repulsion is thus the strongest for the t_{2g} orbitals which, in this case, have the highest energy, while the e_g orbitals have the lowest energy. In other words, the splitting of the d orbitals happens exactly in the opposite order, when compared to an octahedral complex.

One of the tremendous advances allowed by crystal theory is the explanation of the colors of aqueous solutions of metal ion complexes (see Figure 6.8). An interesting experiment involves chromium (III) salts. Let us start from a green solid salt, the chromium (III) chloride hexahydrate $CrCl_3 \cdot 6H_2O$, and dissolve it into water. Initially, the dissolution gives an aqueous solution of $[CrCl_2(H_2O)_4]^+$ that is green. Then, as time goes by, the solution becomes violet, as a ligand exchange reaction takes place and gives rise to the $[Cr(H_2O)_6]^{3+}$ complex. Essentially what happens here is that the two Cl^- ligands are replaced by two more water molecules giving rise to a perfectly octahedral metal ion complex. This means that the types of ligands that surround the same metal ion can result in different colors. Similarly, while $[Cr(H_2O)_6]^{3+}$ is violet in solution, the chromium (III) complex $[Cr(NH_3)_6]^{3+}$ gives a yellow solution. Crystal field theory can explain this difference in color between the two solutions. The chromium (III) ion has three electrons on its d orbitals, and when it is involved in an octahedral ion complex, the electron configuration is $(t_{2g})^3 (e_g)^0$. Thus, if the complex is subjected to light, it can absorb radiation that triggers an electronic transition with one electron moving from t_{2g} to e_g. The wavelength for the absorbed light corresponds to the energy of the electronic transition. In the case of $[Cr(H_2O)_6]^{3+}$, the green-yellow color is absorbed (around 500 nm), meaning that we only see the complementary color, and thus a violet solution. Now for $[Cr(NH_3)_6]^{3+}$ the blue-violet color is absorbed which means that we observe the complementary color and thus a yellow solution. Since the wavelength for blue is smaller than for green, the energy associated with a blue light is greater than for a green light. In other words, the light absorbed by $[Cr(NH_3)_6]^{3+}$ is more energetic than that absorbed by $[Cr(H_2O)_6]^{3+}$or, differently put, the energy difference involved in the splitting of t_{2g} and e_g is greater in the case of $[Cr(NH_3)_6]^{3+}$. This means that the ligand NH_3 creates a stronger crystal field than H_2O. It has been found that the order of increasing ligand field strength is $I^- < Br^- < Cl^- < F^- < H_2O < NH_3 < NO_2^- < Cl^-$ (spectrochemical series).

	Fe (II)	Fe (III)	Co (II)	Cu (II)	Al (III)	Cr (III)
Hydrated ion	$[Fe(H_2O)_6]^{2+}$	$[Fe(H_2O)_6]^{3+}$	$[Co(H_2O)_6]^{2+}$	$[Cu(H_2O)_6]^{2+}$	$[Al(H_2O)_6]^{3+}$	$[Cr(H_2O)_6]^{3+}$
	Green	Yellow/brown	Pink	Blue	Colorless	Green
With OH⁻	$[Fe(H_2O)_4(OH)_2]$	$[Fe(H_2O)_3(OH)_3]$	$[Co(H_2O)_4(OH)_2]$	$[Cu(H_2O)_4(OH)_2]$	$[Al(H_2O)_3(OH)_3]$	$[Cr(H_2O)_3(OH)_3]$
	Green	Brown	Blue	Blue	Colorless	Green

FIGURE 6.8   Colors of various transition metal complexes in solution.

This has several important consequences for the magnetic properties of ion complexes since it affects their number of unpaired electrons or spin number. Indeed, two choices can be made if there are between 4 and 7 d electrons: either putting as many as possible into the low energy t_{2g} orbital or distributing them to obtain a maximum number of parallel spins. Let us compare for instance two complexes obtained from the iron (III) ion. The Fe^{3+} ion has five valence electrons on the d orbitals and gives a high spin complex $[Fe(H_2O)_6]^{3+}$, while it gives a low spin complex with $[Fe(CN)_6]^{3-}$. This result can be explained by the strength of the crystal field which increases (see Figure 6.9) when the ligands H_2O are replaced by CN^-. Indeed, when the energy difference between t_{2g} and e_g is small (H_2O case), then the electron configuration for the complex is $(t_{2g})^3 (e_g)^2$ with all five spins parallel to each other (Hund's rule) and thus five unpaired electrons. On the other hand, when the energy difference is large, then the electron configuration is $(t_{2g})^5 (e_g)^0$ and there is only a single unpaired electron. For this reason, $[Fe(H_2O)_6]^{3+}$ is called high spin and $[Fe(CN)_6]^{3-}$ is called low spin. This has a direct effect on their magnetic properties as the magnetic susceptibility increases with the spin.

Let us add that another theory has also been developed by Griffith and Orgel to account for bonding in metal complexes. This involves considering the molecular orbitals of both the metal ion and the ligands. The ligand field theory goes beyond that the crystal field theory as it provides a mathematical explanation for the order in which the ligands appear in the spectrochemical series. Indeed, when using molecular orbitals, one can tell the type of mixing occurring between the orbitals of the metal and of the ligands!

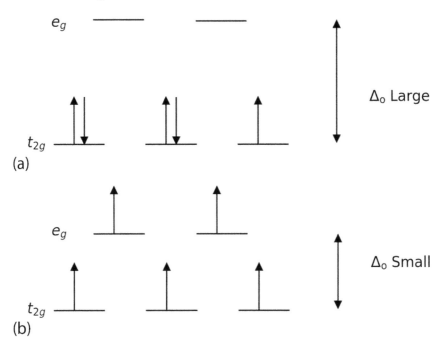

FIGURE 6.9   Spin and magnetic properties: (a) low spin complex and (b) high spin complex.

THE DEVELOPMENT OF CATALYSIS

Catalysis starts to really take off at the start of the 20th century with the discoveries by Arrhenius and Ostwald. Indeed, Arrhenius studies the rate of chemical reactions and finds that the rate constant k is related to the activation energy E_A, the gas constant R and the temperature T through the following equation $k = A \exp(-E_A/(RT))$, in which A is a constant for a given reaction. This relation suggests two ways of increasing the rate of a chemical reaction. The first possible route is to increase the temperature T, which in turn, decreases the ratio $E_A/(RT)$ and thus increases k. The second route is to decrease the value of the activation energy E_A. The activation energy measures the amount of energy necessary to form the reaction intermediate from the reactants. If the conditions of the reaction are modified in such a way that the formation of the reaction intermediate requires less energy, this should lead to a lower E_A and thus to a greater k. Therefore, if a substance foreign to the reaction is introduced and modifies the intermediate, this substance can increase the kinetics of the reaction and become a catalyst for the reaction. Ostwald is the first to identify the importance of catalysts for a wide range of chemical reactions and their numerous applications in organic chemistry and biology as well as in industry. He also proposes a definition for a catalyst: "Based precisely on the participation of the catalyst in the reaction actually occurring, in the sum of which, however, the catalyst is not directly involved, although the partial reactions contain the catalyst as a major chemical component of the process". He states that catalysis relies on intermediate reactions involving the catalyst and concludes: "It must be conceded that no other equally effective principle has hitherto been found in the theory of catalysis". For instance, if we think of a reaction between two reactants X and Y resulting in the formation of a product P, the reaction can be written as $X + Y \rightarrow P$. If we now use a catalyst C that forms a reaction intermediate with X, we then have the following two-step reaction: $X + C \rightarrow XC$ and $XC + Y \rightarrow P + C$. During the second step, the catalyst C is released, meaning that, if we add up the two steps to obtain the overall reaction, we obtain $X + Y \rightarrow P$, which is exactly the same equation as in the reaction using no catalyst! The difference between the two possible pathways (with or without catalyst) is the kinetics. Without a catalyst, the reaction rate is equal to $v = k[X][Y]$. On the other hand, when a catalyst is used, the reaction rate is given by the slowest of the two steps. This is generally the first step and thus $v = k_1[X][C]$. Since we use a catalyst, the activation energy will be lower and thus k_1 will be larger than k! Furthermore, since the catalyst is neither produced nor consumed during the reaction, [C] is constant and thus the reaction rate can be written as $v = k_1'[X]$, meaning that it resembles a first-order reaction with respect to [X], with $k_1' = k_1[C]$.

In 1912, Sabatier (1854–1941) receives the Nobel Prize in chemistry "for his method of hydrogenating organic compounds in the presence of finely disintegrated metals whereby the progress of organic chemistry has been greatly advanced in recent years". Indeed, Sabatier directs a mixture of ethylene ($H_2C=CH_2$) and hydrogen (H_2) onto a column of nickel (Ni) and notices that ethylene is changed into ethane ($H_3C–CH_3$)! The chemical reaction can be written as $H_2C=CH_2 + H_2 \rightarrow H_3C–CH_3$. Sabatier concludes: "Thus, nickel appeared to us to possess a remarkable capacity to hydrogenate ethylene without itself being visibly modified, i.e. by acting as a catalyst". He then decides to use the properties of nickel to hydrogenate acetylene (HC≡CH) and finds that it converts into ethane in the presence of nickel according to the following reaction: $HC≡CH + 2H_2 \rightarrow H_3C–CH_3$. He also manages to hydrogenate benzene (C_6H_6) into cyclohexane (C_6H_{12}) still in the presence of nickel and at a temperature of 180°C according to $C_6H_6 + 3H_2 \rightarrow C_6H_{12}$. This leads him to formulate in 1901 the experimental conditions for a hydrogenation method that works on many compounds: "Vapor of the substance together with an excess of hydrogen is directed on to freshly reduced nickel held at a suitable temperature (generally between 150 and 200°C)". This is the birth of what is now known as heterogeneous catalysis! Here heterogeneous means that the reactants are in a different phase (gas phase) compared to the metal (solid phase). Sabatier wonders: what is the mechanism of the catalyzed reaction of hydrogenation? He quickly comes up with a hypothesis: "I assume that hydrogen acts upon the metal by very rapidly producing a compound on its surface". He identifies this compound as an hydride and explains that: "The hydride thus produced is readily and rapidly dissociated, and if it is placed in the presence of substances capable of using hydrogen it gives it up to them, at the same time regenerating the metal, which again produces the same effect, and so on". This allows him to understand the level of activity of the nickel catalyst towards hydrogenation. For example, he finds that pure nickel is very active and easily hydrogenates benzene. He explains this result by invoking the formation of the perhydride (NiH_2) reaction intermediate. However, if the nickel contains impurities, it becomes less active because the reaction intermediate is this time Ni_2H_2 which according to him is "a poorer hydride [...] incapable of reacting with benzene, though active in respect of ethylenic hydrocarbons or nitro derivatives". Sabatier also understands that the formation of these reaction intermediates and the capacity of metals in fixing hydrogen means that metals can catalyze not only hydrogenation reactions but also dehydrogenation reactions! He states: "Thus the concept of a temporary compound prompted me to use finely divided metals, first as hydrogenation catalysts and then as dehydrogenation catalysts". He verifies this by using powdered copper at 250°C as the catalyst and finds that primary alcohols ($R–CH_2OH$) are dehydrogenated into aldehydes (RCHO) and secondary alcohols ($R_2–CHOH$) into ketones (R_2CO)! The chemical equations can be written as $R–CH_2OH \rightarrow RCHO + H_2$ and $R_2–CHOH \rightarrow R_2CO + H_2$.

Catalysis can also occur within a single phase. In this case, when, for instance, all reactants and catalysts are in an aqueous solution, the catalysis is said to be homogeneous. This is what happens very often in nature, as in the example of enzyme catalysis. Enzymes are biological molecules, in most cases proteins, that catalyze reactions essential to living organisms. For example, ATP synthase is the enzyme that catalyzes the production of ATP molecules for energy storage. Another example is trypsin which catalyzes the cleaving of proteins in the body for digestive purposes. Let us add that Nature has often been a source of inspiration for chemists. Optically active or chiral molecules are often created during natural and biochemical processes, such as, for instance, as noted by Knowles (1917–2012), monosodium L-glutamate, L-lysine and L-mentol. This has prompted scientists to analyze these biochemical syntheses and to develop new optically active reactions to obtain their enantiomers, including D-amino acids. This new discovery leads to the award of the

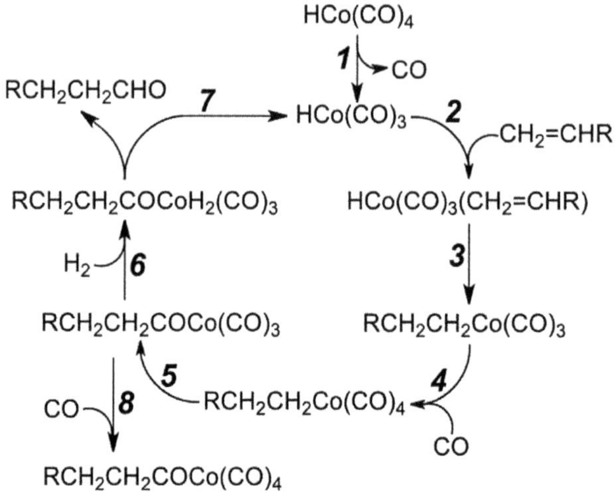

FIGURE 6.10 Hydroformylation reaction.

Nobel Prize in chemistry in 2001 to Knowles and Noyori (1938–) "for their work on chirally catalyzed hydrogenation reactions" and to Sharpless (1941–) "for his work on chirally catalyzed oxidation reactions". Finally, transition metal salts can also be used as homogenous catalysts. This is the case of the cobalt-catalyzed hydro-formylation reaction discovered by Heck (1931–2015) and Breslow (1931–2017) in the 1960s, now replaced by a rhodium-catalyzed reaction (see Figure 6.10). Let us add that the Heck, Negishi (1935–) and Suzuki (1930–) receive in 2010 the Nobel Prize in Chemistry "for palladium-catalyzed cross couplings in organic synthesis". These reactions have made it possible to synthesize complex molecules like dis-codermolide. This molecule is a drug used to attack cancer cells. Discodermolide is a marine natural product that can be isolated from a species of sponge living in the Caribbean Sea. Unfortunately, there is very little of discodermolide available in nature, which makes its synthesis in the lab a great achievement.

ADSORPTION: FROM PHYSISORPTION TO CHEMISORPTION

Heterogenous systems have intrigued scientists for many centuries. For instance, in his book "*Idiota de Staticis Experimentis*", Cusanus (1401–1463) writes in 1450: "If anyone hangs dry wool at one side of a big balance and loads the other side with stones until equilibrium is established, at a place and in air of moderate temperature, he would observe that the weight of the wool increases with increasing humidity and decreases with increasing dryness of the air. By these differences it is possible to weigh the air, and one could perhaps perform weather forecasting." In fact, Cusanus realizes that the humidity of the air, or in other words, the water it contains, can be measured through its ability to stick to a solid material (dry wool) or, as we now call it, to adsorb.

Over the next centuries, scientists develop more and more accurate methods to measure the amount of gas that can be adsorbed by a variety of solid materials. Let us mention the adsorption experiments of Scheele and Priestley on charcoal during the 18th century, as well as the design of the first accurate adsorption balance by Warburg and Ihmory in 1886. Indeed, it is around the end of the 19th century that Kayser invents the term adsorption and plots his results in what is now called an adsorption isotherm. To do so, he reports the volume of gas adsorbed against the pressure of the gas. Indeed, according to Kayser, what occurs is "Adsorption at a surface which is different from absorption in the interior".

However, a question remains: how can this phenomenon be explained? The beginning of an answer comes at the turn of the 20th century with van der Waals. Indeed, he understands that, at the boundary between two phases, there is not really a sharp change from one phase to the other. Instead, there is a transition layer that develops between the two phases. Scientists, like Eucken and Bakker, are inspired by this finding and apply it to gas adsorption. For instance, Eucken characterizes the boundary between the solid material and the gas as a transition layer, very much like a miniature atmosphere, in which the gas is attracted to the surface. Langmuir is the first to develop a theory for gas adsorption that is confirmed by many experiments. Langmuir observes: "I was led to believe that when gas molecules impinge against any solid or liquid surface, they do not in general rebound elastically, but condense on the surface, being held by the field of force of the surface atoms". He also understands that adsorption arises from this balance between the phenomena of condensation and evaporation at the solid surface! Langmuir goes on to characterize the adsorbed phase (see Figure 6.11): he understands that, when the surface forces are strong, there is condensation and the gas molecules form a layer on the solid surface. On the contrary, when the surface forces are weak, the gas molecules do not stick to the surface and evaporate, which means that a very small fraction of the surface is covered. Furthermore, he has the intuition that, because of the short range of the surface forces, the molecules within the adsorbed layer "orient themselves in definite ways in the surface layer since they are held to the surface by forces acting between the surface and particular atoms or groups of atoms in the adsorbed molecule". Considering thin sheets of mica as the solid material, he posits that the crystal surface contains a finite number of what he calls "elementary spaces" that can hold one adsorbed molecule or atom. Indeed, he represents the mica crystal surface as a regular lattice of oxygen and hydrogen atoms which, depending on their arrangements, constitute different elementary spaces. This allows him to understand why adsorption often takes place in steps, when, for instance, all elementary spaces of type I are completely covered, before the elementary spaces of type II start to adsorb additional gas molecules. Langmuir's idea for his theory is the following. He begins by examining the rate at which molecules condense at the surface. For this,

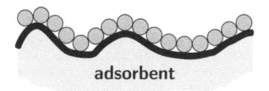

FIGURE 6.11   Langmuir's model for adsorption.

he understands that only a fraction α striking the surface at a velocity μ will adsorb on the bare surface. He denotes by θ the fraction of the surface that is bare or, differently put, not already covered by molecules of the gas. This gives him the rate of condensation on the surface as the product $\alpha\mu\theta$. He also analyzes the rate at which the molecules evaporate from the surface as $\nu\theta_1$ in which ν is the rate at which the gas would evaporate if the surface were completely covered and θ_1 is the fraction actually covered by the adsorbed molecules. Of course, we have $\theta+\theta_1 = 1$, since the entire surface is either bare or covered by molecules. He identifies that, at equilibrium, the two rates are equal, meaning that $\alpha\mu\theta = \nu\theta_1$. He plugs $\theta+\theta_1 = 1$ in this equation to obtain $\theta_1 = \alpha\mu/(\nu + \alpha\mu)$. He then defines σ as the relative life of a molecule adsorbed on the surface with $\sigma=\alpha/\nu$ which leads him to write that $\theta_1=\sigma\mu/(1+ \sigma\mu)$. This allows Langmuir to recover the behavior observed in adsorption experiments. For instance, at low pressure, $\sigma\mu$ is very small since μ is proportional to the pressure. This means that $\theta_1 = \sigma\mu$. Indeed, the denominator $(1 + \sigma\mu)$ is very close to 1 since $\sigma\mu$ is very small ($\sigma\mu \ll 1$). And as shown by experiments, the amount adsorbed is proportional to the pressure. Then, at high pressure, $\sigma\mu$ is very large, giving the following result: $\theta_1 = \sigma\mu/(\sigma\mu) = 1$. In this case, the denominator $(1 + \sigma\mu)$ is very close to $\sigma\mu$ since $\sigma\mu$ is very large ($\sigma\mu \gg 1$). This implies that at high pressure, the entire surface is covered with molecules. Finally, let us add that Langmuir recognizes that the type of surface is very important in the adsorption process. In particular, molecules may be more strongly adsorbed on metallic surfaces, while the same molecules are weakly adsorbed on surfaces like mica or glass. This is what we now know as the difference between chemisorption and physisorption.

Chemisorption refers to the formation of bonds between the adsorbed molecules and the metal atoms of the surface. In the case of physisorption, there is no chemical bonds formation (the interaction with the surface atoms is much weaker). Chemisorption is what Sabatier (see Figure 6.12) had postulated for the reactions of the hydrogenation of alkenes (for example, ethylene) on nickel columns. Indeed, experiments in the 1940s show that when ethylene undergoes chemisorption on a catalyst, its double bond becomes "open", meaning that an "attachment" takes place between the carbon atoms and the metal atoms of the catalyst. They establish this through an exchange reaction between ethylene and a heavy hydrogen atom known as deuterium (D^2). Deuterium is an isotope of hydrogen with a nucleus composed of one proton and one neutron. The opening of the double bond coupled with the dissociative chemisorption of H_2 molecules on Ni paves the way for the formation and release of ethane molecules (H_3C-CH_3). This reaction mechanism is later proved to be correct by experiments, known as high-resolution electron energy loss spectra of adsorbed molecules, and by quantum chemistry calculations.

Let us add that, nowadays, zeolites are a class of porous materials that are extensively used in industry. Their origin dates to 1756 when the first zeolite is discovered by the Swedish mineralogist Cronstedt. He notices that zeolites swell up when they are heated, giving them their name from the Greek ζέω: *zeo* = to boil and λίθοσ: *lithos* = stone. Scientists quickly realize that these minerals adsorb a wide range of liquids including water, alcohols, benzene and chloroform. Their structure is first postulated to be composed of open spongy frameworks by Friedel. It is later determined through X-ray diffraction by Taylor and Pauling. McBain understands that

FIGURE 6.12 Paul Sabatier (1854–1941).

they can separate molecules based on their sizes and thus act as sieves, leading to the term "molecular sieves". Since then, synthetic zeolites are increasingly developed for applications such as fluid catalytic cracking catalysts for the petrochemical industry, separating O_2 from N_2 in air separation processes and, more recently, in environmentally driven applications, for example for the removal and recovery of volatile organic compounds.

THE DISCOVERY OF POLYMORPHISM

By the end of the 18th century, the concept of crystal structure is known from the work of Haüy. And each chemical substance is thought to exhibit only one crystal structure. However, the work of a German chemist and mineralogist Klaproth (1743–1817) raises some doubts on the latter. Indeed, he notices around 1800 that the crystal of calcium carbonate ($CaCO_3$), also known as calcite, has the same chemical composition as a mineral found in Spain and a different crystal structure! The second $CaCO_3$ crystal is called aragonite, because it is found in Molina de Aragón (Castilla la Mancha, Spain). Let us add that Aragón is also the name of a Spanish province. This amazing result is obviously at odds with the idea, valid at the time, that a compound cannot exist in more than one crystalline form. Scientists try to understand Klaproth's strange observation and think, at first, that the two crystal forms are due to the presence of different impurities or, in other words, to the presence of magnesium (Mg) or iron (Fe) in calcite and of strontium (Sr) in aragonite. Today, it is known that aragonite and calcite are indeed two crystal structures of $CaCO_3$. They can be observed very easily in seashells, with calcite constituting the middle layer (or prismatic layer) of the seashell, while the inner layer of nacre is made of aragonite.

Mitscherlich (1836–1918) works in Berzelius' laboratory in Stockholm, when he makes a very interesting discovery on the arsenates and phosphates. He finds that sodium dihydrogen phosphate ($NaH_2PO_4\,H_2O$), while very similar to sodium dihydrogen arsenate ($Na_2H_2AsO_4\,H_2O$), does not crystallize in the same form. Of course, he starts by thinking that there is some impurity in the crystal. He therefore carries out again the same experiment many times. By changing the experimental conditions, he manages to obtain two possible crystal structures for $NaH_2PO_4\,H_2O$. The first is different from that observed for $Na_2H_2AsO_4\,H_2O$ and the second is identical to it! He calls this new phenomenon "dimorphism".

In the next few years, this concept gains in generality as not only minerals or inorganic compounds, but also organic molecules, are found to exhibit more than one crystal form. For instance, Wöhler and Liebig find in 1832 that benzamide (C_7H_7NO) can also be dimorphic. Furthermore, in 1841, Berzelius (see Figure 6.13) has the intuition that elements can also be dimorphic. Indeed, studying two samples of solid sulfur, he identifies strong differences between the two samples and finds that no theory can account for these differences. He proposes a new term to denote this new phenomenon: the two forms of sulfur are called allotropes from the Greek ἄλλοσ: *allos* = other and τρόποσ: *tropos* = turn (literally another turn for the same substance or, in other words, another behavior). Allotropism is also found to apply to phosphorus with the two allotropes being of different colors: the famous red phosphorus and

FIGURE 6.13   Jöns Jacob Berzelius (1779–1848).

the well-known white phosphorous. Finally, another notable example of allotropes is the case of carbon that exists either as graphite under ambient conditions, or as diamond at high pressure. Of course, a diamond generated at high pressure deep under the surface of the Earth takes a very long time to convert into graphite. This is the reason why we can observe diamonds every day. This very long timescale for the conversion is called metastability. Differently put, diamond is metastable under ambient conditions, whereas graphite is stable for these conditions.

Another breakthrough is the discovery by Lehmann (1855–1922) in 1877 that compounds identified as dimorphic can transform into one another by simply being put in contact with one another. He also understands that temperature plays a role in determining the relative stability of the two crystalline forms. For some compounds, one form remains the stable form for all temperatures, defining this class of materials as monotropic. For others, the relative stability changes when a given temperature is reached, leading to enantiotropic materials. This prompts the famous Ostwald to develop a general framework for such transformations. Let us add that, at that point, it is known that compounds can exist in more than two crystalline forms. Hence the word dimorphic is replaced by polymorphic and the phenomenon is now called polymorphism. One of Ostwald's long-lasting achievements is the formulation of the "step rule" also known as the "law of successive reactions". In 1897, Ostwald explains that: "On leaving any state, and passing into a more stable one, that which is selected is not the most stable one under the existing conditions, but the nearest". For instance, during crystallization, the liquid gradually converts into a first metastable polymorph, which then itself transforms into a second, until the stable polymorph is obtained. Ostwald also clarifies the relation between polymorphism and allotropy. In 1912, he writes that there "is really no reason for making this distinction, and it is preferable to allow the second less common name (allotropy) to die out". From then on, scientists like Tammann and Burger make extensive use of thermodynamics to classify and rank polymorphs according to their properties, such as the heat of transition between polymorphs, the heat of fusion and their entropies of fusion.

However, a definite proof of polymorphism is only made possible by the advent of X-ray crystallography. In 1938, Robertson and Ubbelohde elucidate the crystal structures of two polymorphs of resorcinol $C_6H_4(OH)_2$. They observe that around 74°C ordinary resorcinol (α-resorcinol) undergoes a transformation into a denser crystalline form (β-resorcinol). They find that the two structures differ in the patterns found for hydrogen bonding between molecules. The concept of polymorphism is now firmly established!

In 1965, McCrone provides the first modern definition of what a polymorph is: "A polymorph is a solid crystalline phase of a given compound resulting from the possibility of at least two different arrangements of the molecules of that compound in the solid state". He also understands that polymorphism occurs extremely frequently: "Every compound has different polymorphic forms and that, in general, the number of forms known for a given compound is proportional to the time and money spent in research on that compound". Indeed, two polymorphs can sometimes form at the same time, leading to the concept of concomitant polymorphism proposed by Bernstein. Molecules of organic compounds can also crystallize with solvent molecules, resulting in the formation of solvates (hydrates in the case of water) or pseudo-polymorphs. Another example, of particular interest to the pharmaceutical industry, is the case of co-crystals defined in 2012 as "Solids that are crystalline single phase materials composed of two or more different molecular and/or ionic compounds, generally in a stoichiometric ratio, and which are neither solvate nor simple salts". Controlling which polymorph is obtained as the outcome of the crystallization process is very important, since polymorphs often have dramatically different physical properties. For instance, the polymorphs of the 5-methyl-[(2-nitrophenyl)amino]-3-thiophenecarbonitrile notoriously display a wide range of colors. Indeed, this compound is also known as ROY, standing for red, orange and yellow, since ROY crystals can look like red prisms, orange needles, yellow needles and orange-red plates.

Polymorphs of the same molecule can have very different solubilities. This is the case of a HIV-I protease inhibitor known as ritonavir. Here, polymorph I is soluble and polymorph II exhibits a very low solubility. In other words, form II has a very low bioavailability and requires a specific formulation to be used as a pharmaceutical drug. Let us add that polymorphism is undoubtedly everywhere. Cocoa butter (2-oleoylpalmitoyl stearin) has six crystalline polymorphs. Of these six polymorphs, only form V tastes good! Indeed, it exhibits a good snap and no blooming unlike the other forms.

POLYMERS AND MACROMOLECULES

Berzelius is yet again at the center of a new discovery. Indeed, he tries to understand why compounds can have the same chemical composition and, at the same time, exhibit very different properties. He calls this phenomenon "isomorphism" (Greek ἴσοσ: *isos* = equal and μορφώ: *morpho* = shape) or, as we now call it, isomerism (Greek ἴσοσ: *isos* = equal and μέροσ: *meros* = parts). First, he notes that "Substances are isomorphous when they possess different properties although their chemical composition and molecular weight are alike" and defines as metamerism (Greek μεσα: *meta* = beyond or changed and μέροσ: *meros* = parts) substances in which "an equal number of the same elements are joined in a different manner". This is the case, for instance, of ethyl formate ($HCOOCH_2CH_3$) and methyl acetate (CH_3COOCH_3). These two molecules have the same formula, $C_3H_6O_2$, but the chemical bonding within the two molecules is very different! He also realizes that there are several other cases in which chemical compounds "have different molecular weights, mostly multiples of each other but the same percentage composition". Indeed, let us compare ethylene (C_2H_4) and butylene (C_4H_8). Both compounds have the same percent composition for carbon and hydrogen, but dramatically different physical properties including very different vapor densities. He calls this other type of isomorphism "polymerism" (Greek πολύσ: *polus* = many and μέροσ: *meros* = parts). The name polymer and the idea that a molecule can be a multiple of smaller entities (monomers) are born!

From that point on, chemists start to understand that polymers play a major role, most notably in Nature. Payen, a French scientist, starts to study plants around 1840 and discovers that the same fibrous substance can be found in many plants and plant materials. He calls this substance "cellulose" (from the French *cellule* = cell and *ose* = sugar) and discusses its properties in *"Mémoires sur les dévelopements des végétaux"* (Reports on the Developments of Plants). He elucidates the chemical formula for cellulose by making it react with acids and ammonia: *"La cellulose, isomérique avec l'amidon, la dextrine et l'inuline, constitue la substance même des parois des cellules vésiculeuses, polyhédriques ou allongées en fibres, tubes, vaisseaux ou trachées; a peu près pure, elle forme les parois cellulaires des spongioles, des perispermes, des fibres textiles, etc"*. (Cellulose, an isomer of starch, dextrin and inulin, constitutes the matter that makes up the walls of vesicular cells, polyhedral or elongated in fibers, tubes, vessels or tracheas; when it is about pure, it makes up the cell walls of spongioles, perisperms, textile fibers, and so on.) Furthermore, he notices that *"la cellulose injectée de matière azotée et de silice forme l'épiderme ou la cuticule épidermique des tiges et des feuilles"* (cellulose injected with nitrogen and silicon matter constitutes the epidermis or the epidermal cuticle of the stems and leaves). The chemical structure of cellulose is now known to be made of D-glucopyranose ($C_6H_{10}O_5$) ring units (see Figure 6.14). The chemical formula of cellulose is indeed a polymer and can be written as $(C_6H_{10}O_5)_n$, where n is the number of times the monomer ($C_6H_{10}O_5$) is repeated. Cellulose is the most abundant biopolymer on Earth. For instance, wood and straw contain about 50% of their weight as cellulose, while cotton contains more than 90 wt%. It has several advantages as a renewable and biodegradable material, with applications in the paper and textile industry, and more recently, as fillers and coating materials.

Inspired by the versatility of natural polymers, chemists start to design experiments in the lab to produce synthetic polymers. A well-known example is the discovery, at the beginning of the 20th century, of Bakelite by an American chemist Baekeland (1863–1944). He builds on experiments in which formaldehyde (CH_2O) is mixed with phenol (C_6H_5OH) to obtain new materials that have a very high mechanic resistance, but are also infusible and insoluble. In 1907, he indicates in his notebook that "All these tests were conducted in concentrated horizontal digester and the apparatus was reasonably tight. Yet, the surface of the blocks of wood does not feel hard although a small part of gum that has oozed out is very hard." He wonders if "the liquid is simply a solution of the hard gum in excess of phenol, then by simple open-air evaporation, I shall be able to accomplish hardening". By the end of the day, he finds a very promising material which is called D and understands that "D is able

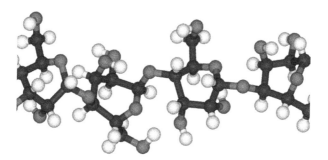

FIGURE 6.14   The biopolymer cellulose.

202

A Mole of Chemistry

to make molded materials either alone or in conjunction with other solid materials as for instance asbestos, casein, zinc oxide, starch [...] and thus make a substitute for celluloid and for hard rubber". D becomes Bakelite and becomes a widely used thermosetting resin because of its inexpensive price and its ability to mold very quickly and to resist extreme conditions of heat and humidity. Examples of applications of Bakelite include iron handles, telephones, washing-machine impellers, automobile distributor caps and many types of insulators.

Another American chemist, Carothers (1896–1937), synthesizes two extremely successful polymers: neoprene in 1932 and nylon in 1938. Neoprene is composed of repeating units of chloroprene $(CH_2ClC=CHCH_2)_n$. It is extremely resistant to wear, and its properties exceed those of natural and synthetic rubber. Nylon is obtained by reacting a diamine $NH_2-(CH_2)_6-NH_2$ (hexametyldiamine) with a diacid HOOC–$(CH_2)_4$–COOH (adipicacid). This means that a diamine with six carbon atoms reacts with an acid composed of six carbon atoms to give the polyamide 6,6 known as nylon (see Figure 6.15). Discovered in 1938, nylon has the same properties as silk, but is much more resistant, leading to its initial use in hosiery. It is reported that, on the first day that nylon stockings are sold in the USA (May 15, 1940), 800,000 pairs are purchased!

Staudinger (1881–1965) is credited with the birth of the field of macromolecular chemistry, which encompasses both natural and synthetic polymers. Around 1920, Staudinger understands that "in common with all organic compounds, the structure of the organic macromolecular compounds [...] involves in addition to carbon atoms, chiefly hydrogen, oxygen, and nitrogen atoms which in accordance with the laws of Kekule's structural theory are bound by chief valences". He realizes that chemical bonding within polymers is the same as in small (regular) molecules. Indeed, he observes that: "the only difference between macromolecules and the small molecules of low molecular substances is one of structural size". He then defines as macromolecules: "substances with a molecular weight greater than 10,000, i.e. the molecules of which consist of 1,000 and more atoms". He also adds: "so far no upper limits can be given for the size of the macromolecules. Macromolecular compounds with a molecular weight of several millions are known." Staudinger is

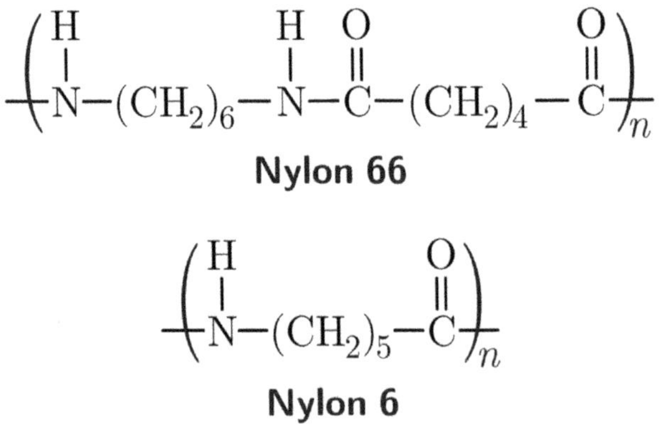

$$\left(-\underset{|}{\overset{H}{N}}-(CH_2)_6-\underset{|}{\overset{H}{N}}-\overset{O}{\overset{||}{C}}-(CH_2)_4-\overset{O}{\overset{||}{C}}-\right)_n$$

Nylon 66

$$\left(-\underset{|}{\overset{H}{N}}-(CH_2)_5-\overset{O}{\overset{||}{C}}-\right)_n$$

Nylon 6

FIGURE 6.15 Molecular structure of nylon.

awarded the Nobel Prize in chemistry in 1953 "for his discoveries in the field of macromolecular chemistry".

Macromolecules also involve molecules that are essential to living organisms such as proteins, enzymes and nucleic acids (DNA, RNA). Proteins are indeed, from a chemical standpoint, made of repeating units of α-amino acids. These compounds have the following general form H_2N-CHR-COOH, and contain an amine function (NH_2), a carboxylic (COOH) functional group and a side chain (R group). The central carbon or α-carbon (connected to R) is chiral, and only the optically active L-amino acids are found in proteins. There are 21 proteinogenic α-amino acids, among which some examples are glycine (in which R=H), alanine (in which R=CH_3) and cysteine (in which R=CH_2SH). The chemical formula of a protein is obtained by creating peptidic bonds between the COOH group of a first amino acid and the NH_2 group of a second amino acid. This gives rise to the primary structure of a protein as, for example, Ala-Gly-Cys-Ala. Proteins are macromolecules with a very long sequence of amino acids, which means that there are ample opportunities for hydrogen bonding within the protein. Such hydrogen bonds occur between hydrogen atoms from the NH_2 group and other electronegative atoms (for instance, O from the COOH group) of an amino acid further down the peptic chain. This leads to the formation of the secondary structure or folding of the protein including the well-known α helix and β sheets. The overall shape of a protein is defined as its tertiary structure and can be determined through X-ray crystallography. Finally, there is also a quaternary structure that describes the number and arrangement of protein subunits. Examples include hemoglobin and DNA polymerase, as well as ion channels.

NANOTECHNOLOGY: EXPLORING THE "ROOM AT THE BOTTOM"

At the start of the 20th century, a new technological revolution begins to take place. Indeed, it is now possible to look at smaller and smaller objects. For instance, Zsigmondy (1865–1929) develops in 1902 what he calls an "ultramicroscope" to determine the size of gold colloidal particles. There are many examples of colloids in nature, starting with the curdling of milk and the clotting of blood. In his Nobel lecture, Zsigmondy describes the following experiment: "I had taken another path which leads me eventually to the preparation of practically equal-sized red gold hydrosols (colloidal gold solutions) by the reduction of gold chloride with formaldehyde in a weakly alkaline solution". He observes that "several of the red gold divisions prepared with formaldehyde [...] appeared perfectly clear in ordinary daylight (like good red wine). They did not settle out their gold, and I was therefore able to call them rightly solutions." He then wonders what the size of these gold particles is. He quickly realizes that these particles are far bigger than an atom as he performs the following test: "The gold particles did not pass through the parchment membrane this showed my gold division. This showed my gold divisions their proper place namely, that they belong to the colloidal solutions." However, a question remains: what is the exact size of these colloidal particles? This is the reason why he invents the ultramicroscope and, together with Siedentopf, manages to determine the size of the gold particles down to a diameter of about ~10 nm (10×10^{-9} m). Zsigmondy receives the Nobel Prize in Chemistry in 1925 "for his demonstration of the heterogeneous nature of colloid solutions and for the methods he used, which have since become fundamental in modern colloid chemistry".

This ability to observe such small objects captures the imagination of scientists during the following decades. Feynman (1918–1988) (Figure 6.16) gives in 1959 a famous talk entitled "There is plenty of room at the bottom". Indeed, he has the following intuition: "I would like to describe a field, in which little has been done, but in which an enormous amount can be done in principle [...] what I want to talk about is the problem of manipulating and controlling things on a small scale". He even formulates the following challenge: "Why cannot we write the entire 24 volumes of the Encyclopaedia Britannica on the head of a pin?" His reasoning is that the head of a pin has a dimension of a 1/16 of an inch and, when magnified by a factor of 25,000 can fit all of the pages of the encyclopedia. To demonstrate that this is feasible, he starts from the resolving power of an human eye (around 1/120 of an inch), divides it by 25,000 and finds the minimal size for a piece of information (a bit) is about 80 Å (80×10^{-10} m) or 8 nm (8×10^{-9} m). Considering an atom of a metal, this means that each bit is roughly 30 atoms across and contains a total of 1,000 atoms to code the information! He concludes: "And it turns out that all of the information in all the books in the world can be written in this form in a cube of material one two-hundredth of an inch wide – which is the barest piece of dust that can be made out by the human eye. So, there is plenty of room at the bottom! Don't tell me about microfilm!" He foresees many intriguing applications, ranging from biology to computer and information science, as well as the development of very small machines able to carry out tasks on an atomic scale. Indeed, quoting Hibbs, he speculates "Although it is a very wild idea, it would be interesting in surgery if you could swallow the surgeon.

FIGURE 6.16   Richard Feynman (1918–1988).

You put the mechanical surgeon inside the blood vessel, and it goes into the heart and 'looks' around [...] it finds out which valve is the faulty one and takes a little knife and slices it out." He also envisions how small machines can assist the human body by carrying out tasks that human organs may not be able to perform properly anymore. This is a vision for what we now call nanomedicine!

A few years later, a Japanese physicist Taniguchi (1912–1999) works on semiconductors and coins the term nanotechnology. In 1974, he states that "nanotechnology mainly consists of the processing of separation, consolidation, and deformation of materials by one atom or one molecule". Yet, the practical realization of devices able to "see" with a resolving power of atomic accuracy has yet to be achieved. A breakthrough comes with the discovery by Binnig (1947–) and Rohrer (1933–2013) of scanning tunneling microscopy (STM). Tunneling is a purely quantum effect. Indeed, in a classical mechanics sense, it is impossible to observe. When a ball rolls along a hill, it cannot go over the top of that hill if it does not have enough kinetic energy to overcome the obstacle. On the other hand, when an electron comes across a potential energy barrier (the equivalent of the hill), there is a nonzero probability for this electron to cross the barrier and go through to the other side. This only happens because an electron is a quantum object! This effect is called tunneling, and it can be observed in molecular compounds like ammonia (NH_3). According to VSEPR, NH_3 is of the type AB_3E, which means that the electron pairs are arranged according to a tetrahedron. This means that, for instance, the top corner is occupied by the lone pair on the N atom and the three corners of the base are occupied by the H atoms. When tunneling takes place, there is a molecular inversion of the molecule, meaning that the lone pair of the N atom flips through N together with the three H atoms. At the end of the 1970s, Binnig and Rohrer decide to work on STM despite their lack of experience in microscopy and surface science: "This probably gave us the courage and light-heartedness to start something which should 'not have worked in principle' as we were so often told". They propose to shape an electrode into a very sharp tip, with just an atom at the end of the tip! Next, they apply a current between this tip and the surface they want to study. They then move very systematically the tip above the surface, hence scanning the surface. What happens here is the tunneling of electrons back and forth between the surface and the tip! This creates a current that allows the surface to be probed with a lateral resolution of atomic dimensions. Binnig and Rohrer are awarded in 1986 the Nobel Prize in physics "for their design of the scanning tunneling microscope". The resolution of STM is indeed extremely impressive with a lateral resolution of 2 Å and a vertical resolution of about 0.1 Å, making this device capable of "seeing" individual atoms and thus probing matter in the greatest possible detail.

The 1980s are marked by the discovery of new nanoobjects and new forms of matters. If we take the seemingly simple case of pure carbon, we already know the two allotropes graphite and diamond. It soon appears that several other types of carbon structures can be found. In 1985, Curl (1933–), Kroto (1939–2016) and Smalley (1943–2005) manage to create fullerenes. Under a helium atmosphere, they subject a carbon surface to a very intense laser pulse which, in turn, vaporizes carbon. Doing so, they obtain strange clusters of carbon atoms sometimes with 60 atoms and sometimes with 70. They quickly discover that clusters of 60 carbon atoms are very abundant and stable. They call them C_{60} or Buckminsterfullerene. This amazing new form of carbon consists of a polyhedron with 20 hexagonal faces and 12 pentagonal faces. It essentially looks very much like a soccer ball! Curl, Kroto and Smalley receive in 1996 the Nobel Prize in chemistry "for their discovery of fullerenes". Let us add that in 1991, Iijima reports the synthesis of new carbon structures shaped as needle-like tubes, the famous carbon nanotubes.

And the story does not end there; a few years later, in 2004, two scientists Geim (1958–) and Novoselov (1974–) find yet another form of carbon, graphene. This time, it is a flat, two-dimensional, atomic crystal in which carbon atoms are arranged in a hexagonal lattice forming a honeycomb. Making this new material is extremely easy as summarized in the "Scotch tape method". In a nutshell, an adhesive tape is applied to a piece of graphite. When the tape is removed, a few layers of graphite stick to the tape. Then, applying the tape to another substrate deposits flakes of graphene on this surface! Graphene has properties that are much superior to other materials: it "is the first 2D atomic crystal ever known to us; the finest object ever obtained; the world's strongest material [...], it is extremely electrically and thermally conductive; very elastic; and impermeable to any molecules". For their discovery, Geim and Novoselov receive in 2010 the Nobel Prize in Physics "for groundbreaking experiments regarding the two-dimensional material graphene".

Index

For Product Safety Concerns and Information please contact our EU
representative GPSR@taylorandfrancis.com
Taylor & Francis Verlag GmbH, Kaufingerstraße 24, 80331 München, Germany

www.ingramcontent.com/pod-product-compliance
Ingram Content Group UK Ltd.
Pitfield, Milton Keynes, MK11 3LW, UK
UKHW020958180425
457613UK00019B/736